The Nightcrawlers

CRITICAL ENVIRONMENTS: NATURE, SCIENCE, AND POLITICS

Edited by Julie Guthman and Rebecca Lave

The Critical Environments series publishes books that explore the political forms of life and the ecologies that emerge from histories of capitalism, militarism, racism, colonialism, and more.

The Nightcrawlers

A STORY OF WORMS, COWS, AND CASH IN THE UNDERGROUND BAIT INDUSTRY

Joshua Steckley

UNIVERSITY OF CALIFORNIA PRESS

University of California Press
Oakland, California

Library of Congress Cataloging-in-Publication Data

Names: Steckley, Joshua, author.
Title: The nightcrawlers : a story of worms, cows, and cash in the underground
    bait industry / Joshua Steckley.
Other titles: Critical environments (Oakland, Calif.) ; 17.
Description: Oakland, California : University of California Press, [2025] |
    Series: Critical environments: nature, science, and politics ; 17 | Includes
    bibliographical references and index.
Identifiers: LCCN 2024052583 (print) | LCCN 2024052584 (ebook) |
    ISBN 9780520413696 (cloth) | ISBN 9780520413702 (paperback) |
    ISBN 9780520413719 (ebook)
Subjects: LCSH: Fishing bait industry—Canada. | Baitworms—Economic
    aspects—Canada.
Classification: LCC SH448 .S75 2025 (print) | LCC SH448 (ebook) |
    DDC 338.3/717509713—dc23/eng/20250219
LC record available at https://lccn.loc.gov/2024052583
LC ebook record available at https://lccn.loc.gov/2024052584

GPSR Authorized Representative: Easy Access System Europe, Mustamäe tee
50, 10621 Tallinn, Estonia, gpsr.requests@easproject.com

34   33   32   31   30   29   28   27   26   25
10   9   8   7   6   5   4   3   2   1

CONTENTS

ILLUSTRATIONS

FIGURES

MAP

TABLES

ACKNOWLEDGMENTS

When I began my PhD, I never imagined I would write about worms in Ontario; I was planning to write about bananas in Haiti. Although life circumstances change, the people who have supported me throughout the past few years have not. I need to begin by thanking my friends and colleagues in Haiti who played a pivotal role in shaping my worldview. Boumba, Jean Remy, and the whole reforestation team, thank you for exemplifying what community-based development can be. I should also thank Mèt Fèyvèt, for having the confidence in me to be a part of your vision for change in Haiti, and Kurt Hildebrand for support and friendship.

This book would also not exist without a timely chat at a London café with Tony Weis, or a summer course taught by Nancy Peluso and Peter Vandergeest. Thank you for these critical discussions about my ever-changing research interests and academic future. My doctoral supervisory committee members Tania Li, Mike Ekers, and Ken Macdonald were extremely helpful in wading through a much longer version of this book. And to my doctoral supervisor Scott Prudham, for your distinctive talent in taking jumbled, half-baked, and unformed student musings and magically turning them into concise and coherent arguments. My postdoc supervisors, Mike Ekers (again!) and Ryan Isakson, as well as the Sustainable Food and Futures Cluster at the University of Toronto Scarborough, which supported me during the writing process: thank you for the flexibility and independence you gave me. At the University of California Press, I want to thank Julie Guthman—an early proponent of this odd worm story—and Kate Marshall, Chad Attenborough, Catherine Osborne, and Jeff Anderson for seamlessly guiding me through this publication process. And, of course, my cousin Jonathan Steckley for his striking cover design.

I want to thank John Reynolds and Alan Tomlin for their expertise and for reviewing the initial drafts of this book. I would also like to thank Hieu Tang, Thana Trakul, and Kaitlyn Lam, who took time out of their busy schedules to help me with their phenomenal interpretation and translation skills. To all the people I spoke with who will not be named, including the coolers, pickers, farmers, advisors, and officials, I truly enjoyed each conversation, and I'm sure each one of you will be able to see how you have informed and shaped this book. The research was supported by a Doctoral Fellowship from the Social Science and Humanities Research Council of Canada.

I also want to thank Jon and Lindsey, Jill and Nathan, and Jenee and Wes, who made their homes available to me during my fieldwork. Also, to my mother-in-law, Rosemary, for tracking down farmer phone numbers. And to my parents, Leigh and Lois, who have always supported me no matter the circumstances (like when upon seeing my Grade 11 report card, they told me, "You know, Josh, it's okay if you don't go to university"). I was fortunate to have grown up in such a loving home.

And finally, to my family. I am proud to have such brilliant and creative kids on the planet. One time, Vienna came home from school and said, "When I grow up, I just want to research and write about things." For a middle-aged father still enrolled at university, I felt validated: "That's exactly what I do!" And Hayden, it is so much fun watching you do what you do, shooting hoops and jamming together on Chili Peppers songs. You both provided a daily reprieve from the heady day-to-day writing. To quote a famous political economist, I have found so much solace in "family life, the children's noise—that microscopic world more interesting than the macroscopic."

Lastly, to Marylynn. Without you, this book would not exist. Take whatever pride you like in that. You are my mentor, my muse, my patron, my partner, and my best friend. I admit you paved the way for our life together, and I will follow you anywhere. To borrow some of the methodological insights gleaned from this book, my ontological status as a human being depends on you. I only exist in relation to you.

Bags Full of Cash

I THOUGHT I SAW A UFO. The cinematic setting was, after all, ripe for an alien abduction. I was a teenager with a freshly minted driver's license in my wallet, driving my parents' Ford Thunderbird all alone out in the middle of nowhere. I had left my girlfriend's house well past midnight and drove along the dirt roads that led me back to New Hamburg, the small southwestern Ontario town where I lived. The roads were empty and dark. There were no streetlights or traffic lights, nor were there any main intersections, corner stores, or flashing stop signs. In a few more hours, the sun would rise and illuminate the picturesque century-old farmhouses, barns, and silos standing high above the rolling acres of budding corn stocks. But in the early morning darkness of that particular night, I could only see the few dozen feet of gravel road quickly passing underneath the car as I drove home.

The mysterious lights appeared deep in a farmer's field. The dozens of flickering lights bobbed asymmetrically up and down but remained close to the soil, as if the wobbly spacecraft was having trouble lifting off. I slowed and rolled down the car window to get a better look, and my imagination came back to earth. It was not a rickety old UFO but just a bunch of people wearing headlights. They were hunched over, heads close to the ground; they appeared to be digging with their bare hands.

The movie genre in my mind switched from sci-fi to crime. What could possibly be buried in the middle of nowhere that a bunch of people would be searching for under the cover of darkness? I was especially fond of those neo-noir movies in the 1990s, like *Fargo* and *A Simple Plan*, where small-town folk—much like myself—inadvertently get caught up with a nefarious cast of characters who are all chasing around a duffel bag full of illicit cash. What

FIGURE 1. Worm pickers as seen from the road. Photo by author.

else would these night-stalkers be doing if not scouring the ground trying to relocate the loot they had secretly buried in a random farmer's field?

As I continued my slow drive, the Thunderbird's headlights illuminated a beat-up box truck parked along the shoulder of the road. I remember a man was standing next to it, smoking a cigarette and staring at my car as I moved toward him. I casually nodded my head to the man while I steadily pressed the gas pedal to pick up speed. I was not interested in getting tangled up in whatever was happening on that field.

. . .

My high school was on the outskirts of my small town. Our football fields bordered fertile agricultural land that smelled like manure in the fall and spring when the nearby farmers returned their animals' nutrients back to their fields. Students came from neighboring small towns or lived on farms themselves. I told one of my friends, a farm kid, about the suspicious nocturnal activities—the lights, the digging, the truck, and what I assumed to be a bag full of money. I expected him to be as riveted as I was. But he was not. Growing up on a farm, he knew exactly who those people were and what they were looking for. There was no buried treasure. No lost bag of cash. Nor were they digging. They were picking.

Worms. Thousands upon thousands of worms.

Dew worms, he explained, came to the surface on cool, misty nights, and the people walking around with headlamps were "worm pickers," uprooting worm after worm that would eventually be packaged in Styrofoam cups and sold to recreational fishermen by the dozen. No midnight crime story, just fish bait. I was a bit disappointed, and my small, sleepy, nothing-ever-happens small town lost its cinematic appeal. My initial curiosity faded, and I never thought about the worm pickers much after that. Why would I?

It is rare, after all, to catch a glimpse of worm pickers working in the fields. You would have to be driving on rural roads in the middle of the night, under the right weather conditions, and fortuitously pass a farmer's field that just happened to be hosting worm pickers on that specific night. I saw these workers a few more times over the next several years, and when I moved away from my hometown, I did not see them again for nearly twenty years.

In the spring of 2017, however, I was driving late at night on the county roads between London and Waterloo, Ontario. My kids were sleeping in the back seat (I eventually married that girlfriend) after a long day visiting family and old friends. The alien abduction setting was identical—the darkness, the farmland, the gravel roads, and those mysterious lights. I saw a couple dozen bobbing lights in the distance and smiled to myself, thinking about the crime story I had concocted decades earlier. But this time, the dull and monotonous night provoked little intrigue. I knew better. Only worms and worm pickers. There was no bag of money. I continued driving in the rural silence for a moment. . . .

Or was there?

The worm pickers were not digging for a literal duffel bag of money, but certainly a bag of money exists for someone, somewhere. Why else have pickers continued to roam the fields that surround small towns in southwestern Ontario? Who are these pickers? Who buys the worms? Who is making all the money, and what are they doing with the profit?

There was little information. I found a few newspaper articles that described the oddities of this unknown industry, often written with tongue-in-cheek headlines. The strange nighttime lights provoked the titles "Night Creatures Not Men from Mars—Just Worm Pickers"[1] and "Human Robins: Fast Worm-Pickers Can Make Good Money."[2] Another alliterative tongue-twisting title aptly described the weather conditions necessary to draw the worms to the surface: "Willing Worms Wait for Warm, Windless Weather."[3] Recent journalists promised "a story to make you squirm,"[4] calling the business a "Slippery Industry."[5]

From these sketches of the industry, I learned something else that further piqued my curiosity and raised more questions. The rural farmland surrounding the small town where I grew up was not just *one of many* places in the world that supplied recreational anglers with dew worms—it was *the place.* Whether you purchase dew worms from a gas station in Ontario cottage country, a convenience store next to Lake Erie, a Walmart in California, a Bass Pro Shop in Florida, or a bait-and-tackle shop in London, England, nearly all were hand-picked from the farmland that stretches between Toronto and Windsor, Ontario, inadvertently giving the region a unique, though largely unknown, distinction: it is the "worm capital of the world."[6]

The only substantive description of the nightcrawler industry in the academic literature comes from a ten-page book chapter written in 1983 by a former Canadian soil scientist named Alan Tomlin, who, naturally, was the first person I contacted when I began my research. He began our interview with a pithy double-entendre that will often appear in this book: "It's an underground industry. All cash. All the way up and down the chain. . . . I had a lot of visits from the CRA [Canada Revenue Agency], and the OPP [Ontario Provincial Police], and the RCMP [Royal Canadian Mounted Police] through the '80s and into the '90s. . . . They started to realize there's a lot of money in it. Several tens of millions of dollars. . . . All unregulated and below the radar of the tax people."

In the 1980s, Tomlin was known as the "worm guy," partly for his respected research on soil and worms, and partly for his rare knowledge of this underground industry. He was contacted by farmers wanting to know how much money they should charge worm pickers to access their land, or what impact removing millions of worms might have on their soil. He also advised worm entrepreneurs looking to cash in on the bait trade and, on two occasions, testified before a provincial court judge to estimate the monetary value of car trunks full of stolen worms (which ended up being more than $5,000).[7]

When I started talking to farmers in the area, that imaginary duffel full of money I had conjured up in my mind all those years ago suddenly materialized. One farmer told me about the thousands of dollars he made renting land to pickers. When I asked how he was paid, he hesitated, probably thinking about the tax implication of what he was about to say, then smiled. "All cash," he responded. "Comes in a bag." Another farmer, whom I will call Scotty, told me he received a bag full of $15,000 in cash twice a year. He didn't ask the person why payments were always in cash; that was none of his business. But, he said, "It has made me wonder, what's all going on with this? Is it fishy?" Two excellent questions and one mediocre pun.

No one really knows. One worm wholesaler, to whom I will refer as Apollo Baits, noted the difficulty of finding accurate figures around the nightcrawler industry—the number of worms picked, the number of pickers, the amount of land, along with dollar amounts exchanged. From my interview transcript:

APOLLO BAITS: It's going to be a problem for you. I don't know the level of substantiation you'll need, but you won't be able to prove much of anything of what's going on.

ME: My objective is not to do a quantitative study.

APOLLO BAITS: Good.

Farmers don't really know the wholesale value of the worms, or how many are removed from an acre of land in a night. Pickers don't know where they are sold, for how much money, or why there is so much demand for these worms. Anglers don't know (or seem to care) where the worms came from, often assuming they are cultivated like composting worms. The worm wholesalers don't always know why some farmers rent their land and others won't. Government agencies don't really know the value of the industry to the local economy. And no one—myself included—has any specific data on the total number of worms picked, the acreage of land rented, or the number of pickers employed. As one workplace tribunal succinctly stated in 1991: "Worm-picking seems to be primarily a cash business with little information recorded or analyzed by any levels of government. Because of these problems, it is difficult to estimate the exact size of the industry."[8]

We can safely assume that the nightcrawler industry is not an economic juggernaut producing billions of dollars in revenue and employing hundreds of thousands of people in Ontario. But it is not inconsequential either. Based on nightcrawler wholesaler estimates and limited trade data, between 500 and 700 million nightcrawler worms are likely picked from Ontario soils per year.[9] The retail price of a dozen worms varies depending on the year and the point of sale, going from a low of $2.80 US at a Bass Pro Shop in Michigan to $6.99 US at an independent North Carolina bait and tackle shop. If we assume a reasonable average of $4 US per dozen, the retail value of Canadian nightcrawlers is likely between $170 million and $230 million US.

Why have so few people heard about this industry generating hundreds of millions of dollars? How did this banal agricultural region become the worm capital of the world? And, to end with the farmer's earlier question, why is this industry underground and why does it seem so fishy?

DRAMATIS PERSONAE (HUMANUM ET NON HUMANUM)

It is hard to describe the machinations of an industry few people know exists. It is a complicated industry involving soil and macrofauna processes (earthworm physiology, rates of reproduction, soil nutrients), agricultural management (reduced tillage, crop rotations, manure, alfalfa, supply management), national and international policies and events (war, refugees, economic recessions), shifting mental conceptions (ideas of leisure time, good bait, soil health) as well as changing weather (rising temperatures, droughts, polar vortexes). There are no clear protagonists or antagonists who are the clear drivers of the industry; it exists through an ensemble cast of human and nonhuman characters who are arranged and rearranged in particular ways to ensure worms are picked, workers are paid, and profit is made. As such, like the first page of a Shakespearian play, I feel it is necessary to introduce the *dramatis personae*—the cast of characters—to ensure the reader has the proper nomenclature to follow the intricacies of this peculiar story.

If I had to choose a protagonist of this story, it would be *Lumbricus terrestris*, colloquially known as the dew worm, common garden worm, lobworm, or nightcrawler. It is the textbook worm species and enjoyed a well-deserved moment of fame when Charles Darwin wrote his final book about its role in soil formation. It is a type of earthworm that creates burrows deep in the soil, coming to the surface at night under specific weather conditions to feed and mate. This behavior has significant repercussions for the structure of the bait industry, shaping it in ways that drastically differ from the commercial production of earthworms like the red wigglers, European nightcrawlers, or African nightcrawlers.[10] *L. terrestris* is most bountiful in undisturbed soils—prairies, pastures, and no-till agriculture systems, where it also provides critical ecological functions by aerating soil, filtering water, and stimulating microbial activity. Its physical characteristics—primarily its length, girth, and scent—also make it the most popular bait worm for recreational freshwater anglers in North America and Europe.

Today, nearly all worms are picked on dairy farms broadly located between Toronto and Windsor. These farms tend to have clayey loam soils with extraordinarily high *L. terrestris* populations thanks to soil type, reduced tillage practices, perennial crops (hay/alfalfa), and heavy rates of manure application. These social and ecological dynamics have inadvertently turned dairy farms into de facto underground nightcrawler factories—a subterranean ecological means of production privately owned by dairy farmers. Though they never intended to be critical actors in this industry, Ontario dairy farmers often receive lucrative cash offers from bait businesses competing for the right to access the valuable macrofauna in their fields, easily making nightcrawlers the most profitable cash crop in the region.

The worms are picked from the farmers' fields by people aptly referred to as *worm pickers*, or sometimes worm harvesters. Worm pickers are the most visible actors in an otherwise invisible industry, and still, most people in southwestern Ontario have never seen a worm picker at work. They work at night on farmers' fields, picking anywhere between 5000 and 20,000 worms a night, depending on the field and weather conditions. The industry has largely relied on immigrant labor, beginning with post–World War II Italian immigrants who resettled in Ontario. Most contemporary pickers are men and women born in Vietnam who came to Canada as refugees in the 1980s following the Vietnam War. Piece rate wages for pickers currently hover around $35 per thousand worms but float throughout the year and vacillate wildly during extreme weather events. Based on aggregate numbers, there are likely between 780 and 1100 pickers, spending about 100 nights per year picking worms, making an average annual income around $22,000.

Each morning after a night of picking, the worm pickers bring the worms back to a customized truck typically owned by the *crew chief* or the *driver*. In general, the crew chief is the person responsible for securing access to land and labor and making sure these two factors of production meet at the right place and the right time. Most crew chiefs are former pickers themselves who have worked their way up the value chain. After a night of picking, worm pickers will dump their worms into containers and stack them in the back of the crew chief's customized worm-picking truck. By 6:00 a.m., the crew chief takes the truck full of worms to a warehouse, most of which are located around Toronto. The worms are then sold by the thousand to a *worm wholesaler*.

Worm wholesalers purchase the worms from the crew chief and store them at a warehouse at low temperatures, between 3°C and 4°C. Hence, wholesalers are more commonly referred to as *coolers*. The coolers dump the worms into white Styrofoam containers, called *flats*, which are then stacked diagonally on pallets in a refrigerated space with fluorescent lights left on. The diagonal stacking ensures sufficient air for the worms, and constant light discourages any curious worms from venturing out of the containers. The low temperature significantly slows their metabolism, which minimizes their consumption of oxygen and peat moss bedding and slows their reproduction. Some coolers directly employ pickers and crew chiefs and sign formalized contracts directly with dairy farmers. Others simply purchase the worms from independent crew chiefs. As both coolers and crew chiefs can be owners of capital and responsible for hiring labor, I will sometimes lump them together as *nightcrawler capitalists*.

From the coolers, the worms are then sold to other *regional wholesalers* or directly to big *retailers* like Bass Pro Shop or Walmart or small bait-and-tackle shops. From these retailers, the nightcrawlers are purchased for $3 to $7 US per dozen, depending on the retail locations, by *freshwater anglers*, who prize the worm for its characteristics that attract a wide array of fish in a wide array of habitats. Nightcrawlers are often cited as North America's most prevalent and popular live bait for recreational anglers.[11] As an avid angler noted in an Iowa opinion page: "If there is a more universal bait for freshwater fishing than nightcrawlers, I have yet to find it."[12]

OPENING A CAN OF WORMS

This cursory sketch of the human and nonhuman actors involved in the nightcrawler commodity chain answers some of the initial *who* and *what* questions about the industry, but it also raises the more interesting questions of *why* and *how*. Why are nightcrawlers the most popular live bait? Why would an angler even pay for worms they could pick up themselves? And if this species of worm is so ubiquitous, why is there only one worm capital of the world, and why is it based around Toronto? Why not cultivate the worms in a commercial setting like other earthworm species? Why do the worm pickers pay dairy farmers such high rents? Why would a farmer allow the removal of such beneficial macrofauna? Why are most worm pickers Vietnamese immigrants? Why are they paid by the piece? Why would they do such an arduous job?

Making money off a living organism with its own quotidian rhythms, physiological processes, ecological functions, and reproduction rates is a complex undertaking. These traits do not always align with and often directly contradict and constrain the capitalist imperative to generate profit and expand production. What needs to happen to turn the nightcrawler—the same banal creature found in gardens and on flooded sidewalks across the continent—into a valuable commodity capable of generating a retail value in the hundreds of millions of dollars? How do the nightcrawler capitalists of southwestern Ontario confront, adapt to, or transform ecological conditions of nightcrawler production—the soils, the moisture, the food supply, the temperature, etc.—over which they have limited control? And what might this tell us about the relationship between capitalism and nature?

After a two-hour interview discussing logistical headaches, physical aches and pains, cutthroat competition, rampant tax fraud, and money laundering, I asked one cooler if his kids were going to follow in his footsteps and take over the family business. He flipped the question back to me. "You have two kids. . . . You see this as a career choice for them? With what you know? And you know more than anybody." I laughed and responded unequivocally, "I wouldn't set foot in this business." The rest of this book will tell you why.

———

Worm Traps and Worm Queens

A WORM TRAP

In 1969, Oscar R. Jackson filed US Patent 3,543,433 for a spring-loaded contraption made of steel, wire mesh, sponge-like backing, guide arms, and an assortment of levers. It was an odd sort of trap: "This invention relates to a trap for collecting worms." The patent does not identify exactly who will use this invention or why anyone would want to trap worms, but Jackson certainly had a specific industry in mind: "Worms have normally been collected manually by worm pickers ... one problem has been that when the picker approaches closely to the worms on the surface of the ground, the worms are able to detect his [sic] presence," and quickly retreat into their burrows. This makes worm picking an "inefficient and difficult procedure" requiring "considerable labour, resulting in higher costs."[1]

The patent notes Jackson's address was in Niagara Falls, Ontario—home to not only the stunning Horseshoe Falls and its gaudy tourist-scape but also an agricultural microclimate that supports a tender-fruit industry of grapes, peaches, cherries, plums, prunes, and pears. At the time of the patent, this area also produced the tender worm commodity. Dew worms—as we will discover in chapter 1—were considered the most useful and adaptable bait for the postwar industrial workforce spending their leisure time out on a lake, alongside a river, or atop a bridge, dropping their baited hooks in the river below. Sure, there were other kinds of baits and lures, but when it was time to catch the next big one, either for supper or for a story, the dew worms wiggled to the forefront. Selling dew worms became surprisingly profitable and suppliers popped up in pockets around North America where *Lumbricus terrestris* were big and abundant. But finding people willing to work through-out the night was difficult and expensive.

I suspect Jackson had seen worm pickers at night crawling over the golf courses, city parks, and fruit orchards, snatching up worms by the canful, and was likely aware of the sums of money in play. In the late 1960s, worm pickers could make $70 per night, equivalent to over $500 in today's dollars. He must have known how the high cost of labor was impacting the bottom line of local bait dealers, and he saw a profitable opportunity. Jackson's worm trap offered nightcrawler capitalists a way to replace those expensive human beings with a bit of capital investment. His trap, he claims, could be made as large as thirty square feet and could be operated by two people who could deposit the trap in any worm-rich field by truck or tractor, catching "worms in a more efficient manner" and "reduc[ing] overall costs."[2]

Unfortunately for Jackson, his invention never caught on. I found no evidence that his trap caught any nightcrawlers on Niagara farmland or orchard. No cooler I spoke with had ever heard about the contraption, let alone wheeled it out to their picking fields. To be fair, this doesn't mean the trap *couldn't* work. Jackson details the intricate mechanical processes and spongey padding necessary to ensure a successful catch. I imagine he conducted some trials and had enough confidence in his worm-catching machine to pay a law firm to file the patent. The problem, I deduce, was not that it *couldn't* catch worms. The problem for this entrepreneur was that his trap could never pick dew worms faster than the primitive technology of a human body, a light, and a couple of cans.

In fact, a quick description of this unknown industry reveals one of the least dynamic, least cooperative, least innovative, least creative, and least capital-intensive industries in the industrialized world. It appears as an exception to what we consider the progression of capitalist development, especially for an industry operating in the heartland of Ontario agriculture and manufacturing. And yet, despite such a lack of innovation, decade after decade, worms are picked, wages are paid, and tens of millions of dollars are made. How is this possible? How do the nightcrawler capitalists continue to produce such surprising profit through such a banal creature? What is it about this peculiar industry that seems so resistant to technological change? Why has no entrepreneur, Oscar Jackson included, been able to revolutionize the production process, raise labor productivity, and gobble up the super profits that would inevitably ensue? Why does it seem stuck in such archaic labor forms—with workers bending over and picking up worms by hand with a tin can?

Why do they tend to be first-generation immigrants, having settled in Canada in the 1980s? Why is the industry underground? Why must they now pick from alfalfa fields owned by dairy farmers? And why is the rent so damn high?

In answering these questions, I suggest we return to our nonhuman protagonist, *L. terrestris*. This lowly worm, going about its business as we walk above its burrows, is a mulish character. It appears somewhat introverted, preferring lots of space while it tunnels, eats, shits, and mates at its own lackadaisical pace. These simplistic attributes are adored by backyard gardeners and temperate zone farmers, as the worms are renowned for their ability to stimulate microbial activity, filter water, and aerate the soil. But for nightcrawler capitalists who want to turn the worm into a profitable commodity for sale in a competitive market, these attributes are frustrating and costly. Without an efficient way to get the worms out of the ground, nightcrawler capitalists must grudgingly adhere to the dictates of the worm. Far from being a simple commodity for sale at a local gas station, *L. terrestris* appears to play a rather significant role in structuring the capitalist bait industry.

Understanding this eccentric industry requires a framework that enables a historical analysis of how living organisms are commodified and, specifically, how the worm's uncontrollable and indeterminant behavior contingently influences capital accumulation, the conditions of production, rent payments, and labor arrangements, often in surprising and counterintuitive ways. Critical agrarian economy is a good starting point. This framework unpacks how capital confronts nature directly in production, with an accented focus on the materiality and the "difference that nature makes" in shaping capital accumulation. I ultimately coalesce these discussions around the concepts of the *subsumption of labor and nature* under capitalism. I suggest the conceptual interplay between the formal and real subsumption of labor and the formal and real subsumption of nature provides a more incisive understanding of how and why capital accumulates in the nightcrawler industry, how nature is commodified and transformed, who or what is in control of the production process, and ultimately how capitalist value is produced and appropriated through an assemblage of human and nonhuman natures.

PROBLEMS OF NATURE

At the end of the nineteenth century, Karl Kautsky asked why capitalism's revolutionizing force—so visible in urban centers—appeared to make fewer inroads into agricultural production. It seemed as though "agriculture does not develop according to the pattern traced by industry: it follows its own laws."[3] His famous "agrarian question" asks why capitalism develops unevenly in agrarian landscapes, creating class relations, labor regimes, and capital investments that don't fit into neat political-economic categories. Critical agrarian researchers have hotly debated the uneven transformation and incorporation of rural producers into capitalist relations, particularly around the persistence and viability of small-scale farming, rural class differentiation, access to land and resources, rural labor markets, and the gendered dynamics of (re)production, as well as multiplicity of agrarian protest.[4] A particular branch of this discussion—what has been called the "ecological agrarian question"—examines the relationship between the expansion of capitalism and the sustainability of ecological systems, as well as how biophysical processes influence capital accumulation.[5] Stemming from Marx's writings on soil dynamics and industrialized agriculture, the concepts of the *metabolic rift* or *biospheric rift* signal how capitalism systematically disembeds humans from their environment and reconstructs biophysical elements and processes as exploitable inputs and resources for accumulating capital, leading to the subjugation and degradation of the environment.[6] At the firm or industry level, other research highlights how ecological dynamics inherent in agricultural production systems influence class formation, labor regimes, and the organization of commodity production. The biophysical elements necessary for agriculture often create numerous obstacles that capital must deal with in situ or find a way to circumvent altogether. Changing weather, precipitation (or the lack thereof), perishability, long production cycles, diseases, pests, and seasonal labor requirements, as well as the fertility of specific pieces of land, can make agriculture a risky and potentially unprofitable capitalist business. The putative "problem of nature" literature sought to understand how such obstacles shape the trajectories of capital accumulation in industries that confront nature directly in the production process—problems that are distinctly different from those of industrialized factory production.[7]

Obstacles for one business, however, may be an opportunity for another, when profitable ventures can be found upstream or downstream from the site of production. Agribusinesses, for example, can "appropriate" aspects of agricultural production by investing in technologies off the farm to later be sold back on the farm, such as farming equipment, fertilizers, pesticides, or seed. Other industries may "substitute"[8] nature-based commodities and processes with industrial processes that are more easily controlled, such as producing margarine instead of butter, using synthetic fibers instead of cotton, or culturing diamonds instead of mining. The financialization of nature can rake in profit through interest, futures, insurance, speculation, and rent.[9]

Seeing such biophysical aspects of production as an obstacle or opportunity, however, risks understanding nature as a passive external element in production—something that capital must act upon, overcome, transform, or avoid altogether. In contrast, the materialist turn in geography[10] refocused attention on the materiality of nature to track how heterogeneous biophysical and other "lively" characteristics are constitutive of and actively shaping our social, economic, and political lives.[11] Nonhuman nature, too, possesses forms of agency and is capable of work. Richard White's environmental history of the Columbia River is emblematic of the dialectical relationship between humans and nature, arguing that "like us, rivers work. They absorb and emit energy. They rearrange the world." He calls the river an "organic machine"—an energy system that intertwines the "natural" and the "social," with each influencing and intruding upon the other to such an extent that it is difficult to see where, or if, these categories begin and end.[12]

L. terrestris is another sort of organic machine. More often, it is called an *ecosystem engineer*—a rather honorific title for a creature that boasts a pair of cerebral ganglia instead of a brain. Aristotle called earthworms "intestines of the earth," which is an apt metaphor considering their tubular, mushy bodies appear designed as an optimal digestion machine. They have also been compared to rudimentary agricultural technologies, referred to as "nature's plow" or "God's mill." Such metaphors convey the earthworm is "working" to transform, shape, or maintain the soil upon which human society depends. They are not insignificant soil dwellers, only acted upon, but are constitutive of power relations through the work/energy they emit. This was even clear to Charles Darwin, who noted, "It may be doubted whether there are many other animals which have played so important a part in the history of the world, as have these lowly organised creatures."[13]

Incorporating the materiality of nature into our analyses of agriculture and other resource geographies signals that commodity production (and capitalism more broadly) depends on uncommodified biophysical entities and processes to produce profit. If we believe people and machines do work, however nebulously defined, then it follows that nature, whether rivers or worms, certainly works too. "Ecological labor"—the extrinsic reproductive metabolic work "necessary for the regeneration and renewal of ecosystems"[14]—is necessary for capitalist production. Milk production, for example, certainly relies on farmers' labor and expensive machinery, but it also relies on photosynthesis necessary to grow feed crops, carbon cation exchange, the cow's metabolism, its ruminant digestion tracts and the micro-biota embedded within, not to mention the pollinators, fungi, worms, water, nematodes, and . . . well, this list could go on and on. For industries that confront nature in the course of production, the living potentialities and vitality of nature become "inscribed" or "embedded" into capital.[15] The "difference that nature makes"[16] is fundamental and constitutive of capital accumulation.

In the bait industry, *L. terrestris* certainly seems to be doing most of the work as well, being responsible for growing, burrowing, eating, and mating without any application of nightcrawler capital. It is freely producing itself as a ready-to-pick commodity, and the nightcrawler capitalists depend on this uncommodified ecological labor to profit. But if we consider that nature "works," we might also ask what appears to be a simple question: Does nature produce value? Who or what is producing value in the nightcrawler industry? Is it the worms, the cows, the pickers, the capitalists?

CAPITALIST DIGESTION MACHINES
AND SURPLUS VALUE

Value theory is a complex and controversial subject, which I do not intend to detail here, but it merits a brief discussion to understand the dynamics of accumulation in the nightcrawler industry. To be clear, I am talking about a specific kind of value—capitalist value. This is distinct from moral values or environmental values. In Marxist terms, the metric of capitalist value is *abstract socially necessary labor time*—essentially the average amount of labor time it takes to produce a commodity completed by the "workers' average degree of skill" in an average stage of scientific and technological development in the production process. When workers are paid a wage, their capacity to labor also becomes a commodity for sale on the market and thus possesses a value. This capacity to labor, however, is a special sort of commodity because it can produce more value than it is worth in the production process. Marx calls this difference between the value of labor-power and the addition value embedded in the commodity "surplus value." The historical and contemporary struggles between labor and capital are struggles over who gets to claim more of the surplus value created in the production process.

Environmentalists may be quick to point out that arguing labor as the source of value ignores nature's contribution to the production process—certainly, they suggest, bees are working and producing just as much value as, if not more than, the beekeeper! A theory of value that ignores nature's contribution to commodity production (as well as the reproduction of all life on the planet) seems to have a rather glaring blind spot. Marxist scholars respond that it is not Marx who ignores the value of nature; it is the capitalist system that instrumentalizes nature to produce capitalist value through the exploitation of labor.[17] This explanation might appear to dissolve nature into labor, but this is what capitalism must do to get a measure of value.[18] Bees do incredibly useful labor, but who gets compensated through the monetary commodity relations, the bee or beekeeper? It is rather difficult to measure nature's contribution to value in a system governed by monetary commodity exchange, which is also why trying to "price" nature through carbon offsets or ecosystem services is so arbitrary and problematic.[19] But simply stating that capitalism treats labor and not nature as the source of value does little to understand how nature is conjoined in the labor process.

Jason Moore builds on other scholars wrestling with these questions by drawing on White's concept of "work/energy" to better conceptualize how nature is imbricated in capitalist value. Moore suggests capitalism depends on what he terms "Cheap Nature," which must be cyclically restored through the identification and exploitation of "commodity frontiers"—"bundles" of uncapitalized work/energy that can be mobilized and appropriated with minimal capital outlay. It is "cheap" because nature (and non-waged human labor) has already done most of the work for it to exist. He cites the mines of Potosí, pre-contact North American soils, old-growth Norwegian forests, Caribbean sugar plantations, and the proletarianization of unwaged labor as examples where capital investments provide exceptional returns because the unwaged work/energy of humans and nature are appropriated as free gifts. Capitalism has survived precisely because labor productivity—the metric of capitalist value– essentially gets more bang for its buck when "cheap nature" can be identified, mobilized, and rechanneled toward capital accumulation.[20] Moore maintains the measure of value as abstract socially necessary labor time, but argues this will depend on the amount of uncommodified work/energy already done by nature. At an ontological level, this means capitalism has a systemic imperative to avoid paying the full costs of the work, energy, and materials that underpin capitalist production—something that feminist scholars have painstakingly pointed out in relation to the socio-ecological conditions necessary for the reproduction of the workforce.[21] Unpaid work, energy, and other "free gifts" of nature are part and parcel of value theory; in this vein, Richard Walker argues for a joint value theory—a "unified measure of labor-nature time," to make clear that "natural forces, both visible and invisible, are at work in every labor process," and therefore "the work performed by nature is already subsumed in the value calculus *because the average labor time includes the socially necessary amounts of natural inputs and natural forces*" (emphasis in original).[22] Put more bluntly, "capital needs nature,"[23] with human *and* nonhuman work/energy at the heart of capitalist value.

Paul Burkett puts the relationality between labor and nature another way: "The subsumption of labor under capital thus entails a subsumption of nature under capital."[24] The word "subsumption," I suggest, is critical. In everyday parlance, to subsume means to place, include, or absorb something into something larger. Burkett is clear: both people and nature are being subsumed and changed under the gravitational pull of capital. But he goes further to claim that labor cannot be subsumed within capital without nature. How and why does this happen, and what does this mean for our understanding of capital/nature relations? For our purposes here, we might ask how and why a common dew worm is subsumed under capitalism, and what impact this has on the worms, the worm pickers, the nightcrawler capitalists, and their ability to capture surplus value.

THE FORMAL AND REAL SUBSUMPTION OF LABOR AND NATURE

Subsumption is about control and design over the production process and how this impacts surplus value. In the appendix to the first volume of *Capital*, Marx introduces the word as an analytical and historical concept to describe how labor is subsumed within the capitalist production process and how this transforms both. He focuses on two forms of subsumption[25] to describe the historically distinct way and degree to which capital *controls* and *designs* the production process in accumulating surplus value. The *formal subsumption of labor* refers to forms of production where capitalists employ wage labor but rely largely on the inherited labor processes controlled by the laborers themselves. The particular skills, knowledge, and technologies used in production come from the employee, be they a seamstress, a locksmith, a forger, or a worm picker. As the capitalists cannot control or shape the productivity of labor in any radical way, they have few options to increase surplus value; perhaps they make production more orderly and less wasteful (like using an LED headlamp instead of a miner's headlamp for picking worms), expand production, employ more workers, or, critically, extend the hours worked in a day. By doing this, the capitalist increases what Marx calls "absolute surplus value"—that is, to put it crudely, increasing profits by employing more workers and making them work longer and more intensely.

But when the capitalist amasses enough capital and invests in science, technology, and new forms of cooperation, things begin to change, and industry can finally "stand on its own two feet," no longer relying on the particular skills of individuals.[26] Large-scale machinery and the application of science in service of productivity often disrupt the "archaic" or "traditional" labor forms to produce entirely new forms of labor that no longer resemble the inherited skills, knowledge, tools, and methods of production, turning the worker into the proverbial "appendage of the machine."[27] Spinners and weavers are no longer spinners and weavers, but wage workers clocking into the factory and pushing a button to make the huge loom move. Marx calls this transformation *the real subsumption of labor under capital*, where control and design over the labor process are firmly in the capitalists' hands. Instead of expanding employment and working longer hours, capitalists can now radically increase labor productivity by investing in fixed capital like machinery and other technologies. Laborers still receive their wages, but the extra value generated by increasing output goes to the capitalists as *relative surplus value*—the value that comes by increasing the productivity of labor via technological and organizational innovation, as opposed to making labor work harder for longer hours.[28]

In examining the diversity of "nature-based industries," William Boyd, W. Scott Prudham, and Rachel Schurman noticed how Marx's analysis of the subsumption of labor can also be applied to nature.[29] Initially, capital must confront the existing biophysical elements and processes much the same way as the human worker. In this framework, the *formal subsumption of nature* under capital treats nonhuman natures as exogenous stocks of raw materials that produce *absolute surplus value* through "extractivist" logics. Much like the capitalist who inherits the traditional labor forms, the capitalist must work with nature as it is found and work around any obstacles that nature throws up, such as the peculiarities of land[30] or the rhythms of biophysical production time.[31] The capitalist's strategy to increase surplus value via the formal subsumption of nature is rudimentary and analogous to extending the working day: more land is put into cultivation, more trees are cut, more fish are caught, more seeds are planted, more worms are picked. Increasing profit margins comes through *extraction* by *widening* or *extending* production into new geographical spaces and increasing the aggregate volume of production.

The *real subsumption of nature under capital*, by contrast, describes how capital takes control and redesigns ecological processes to increase profit by increasing *the productivity of nature itself.* Industries operating through the real subsumption of nature tend to be in areas of biological commodity production that are more amenable to transformation through genetic modification, breeding practices, antibiotics, and fixed capital investment. This has been shown, for example, in trees,[32] seeds,[33] strawberries,[34] and livestock.[35] Boyd's analysis of broiler chickens is emblematic of how biological production can be accelerated and intensified with genetic improvement via selective breeding, confined feeding, nutritional supplements, and antibiotics.[36] The real subsumption of nature is therefore not simply about making labor more productive but also making "systematic increases . . . in or intensification of biological productivity (i.e., yield, turnover time, metabolism, photosynthetic efficiency)."[37] Whereas the formal subsumption of nature follows an extractivist logic that exploits stocks of resources *extensively*, the real subsumption of nature follows an *intensive* or "cultivating logic," where nature itself is made and "(re)made to work harder, faster, and better."[38]

However, seeing nature as working "harder, faster and better" returns us to the question of who or what is doing the work, and who or what is producing the value. Is a high-yielding hybrid seed really working harder and faster by itself, or might it be forms of human labor having spent countless hours— now embedded in the seed—selectively breeding plants for specific characteristics? Is a chicken working harder and faster today by growing four times larger on its own? Or is it the humans conducting genetic research, developing feeds, and new antibiotics? This was Neil Smith's primary criticism of the subsumption-of-nature concept: it minimizes this centrality of human labor.[39] Social labor (not nature) is the "fulcrum of the production of nature" and the productivity gains that go along with it. It is not that *nature* is working "harder, faster, and better"; instead, the increases in productivity are from the science, technology, and cooperative *social labor* embedded in biophysical processes.[40]

This is undoubtedly true. Hybrid seeds and contemporary broiler chickens do not exist without the massive investment and scientific labor time necessary to achieve such productivity increases. Nature's increased productivity is mediated through labor. This is clear in the analyses around the "production of nature"[41] or the production of "landscapes" where "reified social relations" embedded in material environment serve to orient the production and circulation of commodities.[42] Humans work to make nature work more.

And yet, as I argue here, the reverse also holds true: the characteristics and functioning of ecological processes (often themselves manipulated, transformed, or hobbled by past human labor) are fundamental to the organization of capitalist production and how labor is articulated within it. We can understand these dynamics more clearly if we see the control and redesign of nature not as analogous to the control and redesign of labor but rather as intricately related to them. Put relationally, the control and design of nature within the production process *shape and are shaped by* the control and design of labor. Other scholars have hinted at this relationship. Wim Carton and Elina Andersson call for a theory that sees the subsumption of labor and the subsumption of nature as dialectically related.[43] Boyd and Prudham, meanwhile, upon reexamining their initial framework, suggest a similar exploration is necessary, noting how increasing control over nature might instigate changes in the control and design of varying labor processes.[44]

What would it mean to suggest, in other words, that control and design over nature is not a parallel process to control and design over labor, but that they are intricately conjoined? Oscar Jackson's worm trap failed to rapidly catch worms. Attempts at cultivating nightcrawlers have left mountains of debt. The inability to increase control and design over earthworm production has, in turn, greatly affected the labor process of worm picking, the conditions of worm picking, which people pick worms, and how they exert power in the workplace. How much of an impact? Consider the "Worm Queen."

SUBSUMPTION CONJOINED

I met a woman at a Tim Hortons restaurant in Toronto. She was in her late thirties, well dressed with a friendly smile and manicured fingernails—which I noticed as she handed me a coffee that she bought for me. She told me through a translator about how she came from Vietnam to Toronto nearly twenty years ago on a student visa and found some work at a Vietnamese restaurant and, later, a candy factory. Both jobs paid the paltry $6.85 minimum wage, and she decided to look for something that paid more. A friend told her about a lucrative, albeit unorthodox, way to make more money in less time. No special skills or language were necessary. A funny-looking truck would pick her up and drive her to a farmer's field in the middle of the night, and from there, all she had to do was pick dew worms as fast as she possibly could.

"I was scared," she recalled, telling me about that first night out. She remembered how her headlamp would flood the ground with light and illuminate thousands of worms coming up from the ground like in a horror film. It was scary, cold, dark, and gross; "I literally cried." Her back and knees ached for weeks, and the worms started to infiltrate her dreams. They were not nightmares—just reliving the repetitive motions of clasping worms and slowly uprooting them from the earth, one after another. "They would circle around my head," she laughed.

After a few more weeks, however, her body adjusted to the rigors of picking, and she steadily increased her productivity. She went from picking a few thousand worms a night to 10,000. After a couple more months, on good nights, she could pick 20,000 worms—about as fast as the fastest pickers in the industry. But she didn't stop there. On optimal picking nights, when the weather conditions align, she claims she can pick 30,000 nightcrawlers in a single night, a figure that other pickers I spoke to confirm. Her fellow worm pickers were astounded at her speed; "she is like a superwoman," one told me later. They eventually gave her a well-deserved nickname—the "Worm Queen." Depending on the price of worms and weather conditions, the Worm Queen could make between $600 and $1,000 on such nights. Not bad at the time for a part-time student. In the winter, she works at a warehouse where she picks orders for shipping for $20 an hour or $150 a day, less than half what she makes on an average night picking. Unsurprisingly, she is also quite efficient at this warehouse work, and her supervisor gives her flexible hours during the worm-picking season to ensure he can still have her working full-time during the winter months. "I am a good asset for the company because I'm so fast. . . . They want to keep me."

In comparing the two jobs, she prefers worm picking. "When I heard about it, I wanted to try because they said it's completely free, and your income is dependent on your ability." I was a little confused and followed up with the question, "What do you mean 'completely free?'" She responded that worm picking "is not like a company where you have a supervisor. You don't have to work during a set time of hours. This job, you work at your pace.

When you're tired, you sleep. Hungry, you eat. And if you work well, you get more pay."

The Worm Queen's expression of freedom was unexpected to me. How is it that a supposedly "low-skilled"—as defined on employment classifications—immigrant laborer can make so much money in a physically demanding job operating in an unregulated economy while simultaneously expressing notions of "freedom"? These are not the working conditions normally associated with freedom and autonomy. What's going on here?

The nightcrawler capitalists are stuck with the worm as they can find it; as a result, they are also stuck with the worm pickers as they can find them. The worms, the soil, and the pickers are simultaneously subsumed into capitalism. Worm pickers cannot be subsumed without the worms, nor the worms without the pickers. But they are only *formally subsumed*. The uneven ability of nightcrawler capitalists to control and design earthworm ecology and reproduction will impact where the worms are found, how capital is invested, how production is organized, and how the pickers are articulated within the production process. The subsumption of nature is conjoined with the subsumption of labor. This is the central argument of this book: the extent to which capital can control and redesign nature will shape how capital accumulates and its control and redesign of labor in the production process.

When they are conceptualized as dialectically conjoined processes, the distinction between human and nonhuman natures, particularly the "Cartesian" idea that humans are external to nature, evaporates. Every purported obstacle, opportunity, or surprise that nonhuman nature presents reshapes the organization of labor and capital, which in turn reshapes nature. The conjoined subsumption of labor and nature shows there is no such thing as human productivity or natural productivity; they only exist in their metabolic relation to one another, inextricably conjoined where new technologies, new modes of production, and new means of cooperation persistently and simultaneously shift human/nature metabolic relations.

A conjoined subsumption framework enables an empirical analysis capable of isolating tendencies in how capital confronts a diversity of natures, how it harnesses (or fails to harness) nature's reproductive capacity, what technologies are deployed, how rents are established, who labors, and how surplus value is produced and appropriated. This goes beyond simply seeing how the biophysical materiality of production "matters," so to speak. Instead, it highlights who or what controls and transforms such materiality, how the materiality of commodity production is intertwined with labor and capital, and how the materiality of nature is not static but dynamically shifting, transforming, and refining—sometimes at the behest of capital, and sometimes not.

The nightcrawler capitalists have limited control and design over the ecological conditions of nightcrawler production. It is "stuck" in the formal subsumption of nature, which helps explain its peculiarities, including the piece-rate pay structures, the punishing labor processes, the minimal fixed capital investment, the complex rent relations, and the logic of surplus value extraction. Take surplus value. Where capital relies on labor and nature *as it is found*, absolute surplus value is produced through extensive and extractivist logics. To increase absolute surplus value, the nightcrawler capitalists must follow the worm (chapters 1 and 2) to landscapes that increase either labor productivity or ecological productivity. With no worm-picking machine set to revolutionize the bait industry in sight, absolute surplus value must be captured extensively by finding more land, more worms, and more pickers.

Subsequently, the ecology of *L. terrestris* influences how land rents and property rights are formed. Ground rent becomes an integral cost for industries that confront nature directly in production as capital cannot fully control the ecological conditions of production. Increasing control over nature through technology, for example, would radically alter production to rely less on the specificities of biophysical elements embedded in a specific piece of land, and structure rent relations more around industrial space, commercial zones, or locational advantages. Further applications of science and technology to nature also mean that rent, under the real subsumption of nature, may have little to do with parcels of fertile land but may rather depend on the enforcement of intellectual property. The nightcrawler capitalists, however, operating through formal subsumption, do not have this option; they must rely upon specific pieces of land and deal with the people who own them (chapter 3).

The formal subsumption of labor and nature will also tend to have low compositions of capital (the ratio of constant capital to labor), often with low barriers to entry, whereas shifting to the real subsumption of labor and nature tends to require increases in fixed capital investment, and therefore poses higher barriers to entry. In the nightcrawler industry, characterized by limited control over nature, the low barriers enable a competitive environment conducive to small capitalists and other forms of self-employment. This also relates to the material visibility and institutional legibility of the industry. The formal subsumption of labor and nature creates opportunities for capital and labor to operate "underground" or off the books (chapter 4). The ability to remain largely invisible creates lower-risk opportunities to increase absolute surplus value by avoiding tax obligations. As investments in fixed capital increase, in contrast, it becomes more difficult for nightcrawler capitalists to operate in unregulated ways.

If we look at labor regimes and pay structures, we see that where capital has less control and design over labor and nature, working conditions tend to be physically demanding, with seasonal, irregular, blurry working hours that must adhere to the in situ ecological processes. Work regimes may also tend to skew towards manual over mental labor in the production process. Payment schemes could be piece rate, hourly, contractual, or dependent on advances; capitalists must often rely on workers from marginalized and racialized labor pools who, for reasons beyond their control, are denied high-wage employment opportunities. As the industry transitions to the real subsumption of nature, the materiality of the production process may lose its distinctive and individualistic features, no longer being reliant on particular manual labor skills or in situ ecological processes. Payment structures move toward wage and salary work, while hours may become more stable, with labor gains achieved through collective organizing. At the same time, transforming nature initiates the bifurcation of labor between hourly wage work and more educated and research-oriented labor focusing on the transformation of nature. The nightcrawler capitalists would like nothing more than to replace the Worm Queen with some kind of technology. But they cannot. This has significant ramifications in shaping the labor process and labor agency in the industry (chapters 5 and 6).

Capital moves and manifests in, around, and through nature (and, of course, the reverse, how nature operates in, around, and through capitalism), allowing us to analyze and make sense of the heterogeneity of industries that must confront nonhuman nature during production. It also helps explain why certain bodies are needed at certain times, why some land (or water) is valuable, why some industries go through rapid technological change and others do not, and why workers are paid by the piece, by contract, a wage, or a salary. Consider industries where the production of similar commodities has vastly different labor regimes. There are significant differences, for example, between the matsutake mushroom hunters described by Tsing and the wage laborers who harvest exotic mushrooms grown indoors.[45] The transition from wild salmon fisheries to salmon farming (aquaculture) transforms male-dominated seafaring fishers into salaried or wage laborers responsible for managing closed-circuit or recirculating aquaculture systems.[46] California berry growers have recently begun swapping strawberries with raspberries, the latter of which possess material characteristics that are slightly less labor-intensive, thereby altering the concrete working conditions and affecting the subsequent availability of labor.[47] This can also apply to nonorganic industries. Diamond mines, for example, require specific scientific labor necessary to locate exploitable mines, access to the land (through politically mediated rents, leases, licenses), as well as a combination of expensive mining technology and available low-skilled labor. The ability to culture diamonds that are physically, chemically, and optically identical to mined diamonds, in contrast, radically reshapes labor regimes where exploiting geographically distinct geological formations is no longer the determining factor in production. And, as I will show, the labor regimes associated with the cultivation of red wiggler bait worms are vastly different from those of nightcrawler production. The "production of nature," in these cases, not only results in socio-ecological landscapes structured to facilitate capital accumulation but simultaneously transforms capital investments, the structure of the labor regimes, and the conditions of working bodies.

Thinking through this conjoined relationship also reemphasizes that the characteristics of production are neither static nor teleological, but dialectical and dynamic. Historical circumstances, labor organizing, state regulation, or technological changes may shift production, opening new avenues of accumulation and forms of production while denying others. The technologies developed through real subsumption meant to increase productivity in one sector may open the door to formal subsumption in another, which in turn instigates the pressure for real subsumption again. For example, Uber produces software algorithms (real subsumption) that allow any person with access to a car and phone to become a taxi driver (formal subsumption), only to begin investing in autonomous cars (real subsumption). I want to stress that the relationship is not unidirectional, with formal subsumption leading to real subsumption; it is instead a conjoined relation whereby capitalist control over ecological processes shapes the conditions of labor and capital investments. Labor is constantly rearticulating itself within production processes that have been transformed through past labor-nature relations.

This conceptual interplay allows us to see how the nightcrawler capitalists are stuck in the formal subsumption of nature, unable to take control and transform nightcrawler biologies and ecologies to increase productivity. As a result, they must pay extraordinarily high rent to dairy farmers to access particular worm-dense ecologies that cannot be easily (or rather, profitably) reproduced. The lack of fixed capital investments and the low barriers to entry also enable much of the industry to operate at the margins of the formal economy. Their ability to increase profits depends on strategies of appropriating absolute surplus value—cobbling together more rented land, more cow manure, and more worm pickers. This is not an easy task.

The uneven ability to increase control and design over the worms has created a topsy-turvy industry that operates under cover of darkness while persistently rearranging certain human and nonhuman elements in ways that produce surplus value. In addition to the ensemble cast of characters outlined above, there will be human guest appearances by Charles Darwin, Michael Ondaatje, and Jimmy Carter's cousin Hugh, as well as nonhuman cameos including the Hudson Bay Company, colonial ship ballast, fermented alfalfa, 2,4-D herbicides, cow manure, climate change, and COVID-19—each of which play a role in the development and current configuration of the formally subsumed industry. A common refrain throughout this book is how the collective agency of capital cannot entirely subsume the forces and conditions of production according to its will and must rely on the human and nonhuman biologies, physiologies, and ecologies available to it. The nightcrawler capitalists must be reactive to ecological niches, dairy farm management practices, immigration policies, daily weather, cow manure, and soil structures over which they have limited influence. They rely on all of these contingent factors to turn a profit. The continued existence of this money-making activity called worm picking thus only makes sense by going beyond the typical political and economic human actors to understand how other human and nonhuman elements are persistently arranged, rearranged, sometimes included, sometimes thrown away. We have to analyze this *ensemble* of human and nonhuman characters in the nightcrawler story as an *assemblage*.

ASSEMBLING NIGHTCRAWLERS

The proliferation of "posthumanist," "more than human," "new materialist," critical animal studies, and Actor-Network-inspired research has done much to decenter traditional ideas of human control over the environment, as well as of capital's putative deterministic drives as the primary forces of historical change. Instead, the generative capacities of nonhuman natures—their uncooperativeness, unruliness, and potentialities—require "non-deterministic notions of nature" with an open ontology of agencies, where complex systems operate "without telos or final cause."[48] Capital—viewed as value in motion—cannot dictate how nonhuman natures are produced, nor can the individual capitalist utilize Promethean ingenuity to completely harness or mitigate the complex and unstable living dynamics that surround them. Timothy Mitchell's account of a malaria outbreak in Egypt is an empirically rich account that firmly rejects any notion of capital's "inner logic" and shows how explanations of historical phenomena must go beyond human and capitalo-centric analyses.[49] Even the lowly dew worm itself is an actant, with Darwin himself noting how *L. terrestris* possesses "small agencies" despite the absence of a brain. Jane Bennett notes how "small agencies" can have an enormous ecological impact, such as forests encroaching on the savanna in Amazonia. For her, this distributed nonhuman agency is critical to understanding how "a series of interconnected parts" is continually "being reworked in accordance with a certain 'freedom of choice' exercised by its actants."[50]

Similar research has made much use of the "assemblage" concept—a descriptive concept that emphasizes detailed and open-ended accounts of associations of heterogeneous elements (institutions, organisms, humans, signs, practices) that are provisionally aligned to maintain a coherent whole.[51] Assemblages are non-teleological, with each element affecting, disrupting, and co-constituting the others. Such heterogeneity opens the social, semiotic, and material components to emerge from their entanglement in the assemblage without a single dominant force dictating its constitution. For Martin Müller there are "no pre-determined hierarchies, and there is no single organizing principle behind assemblages ('it is never filiations ... these are not successions, lines of descent'), be it capital or military might. All entities—humans, animals, things, and matters—have the same ontological status to start with."[52]

This is the potential benefit of assemblage thinking. It radically decenters humans as the primary actors determining socio-ecological history and geographies and highlights how these temporally aligned, heterogeneous elements exist in their relation to one another. It not only makes sense of chaotic movement and rapid transformations but also documents how such disparate elements can maintain stability over time. Incorporating nonhuman actors, institutions, and processes into our analysis and methodology permits us to notice things that are typically invisible, ignored, or minimized due to preconceived theories of agents of change and ontological drivers of socio-ecological change. Capital and humans alike do not preexist or represent the determinant agents but are constituted by and through their contingent and temporal relationship to other elements in the assemblage.

The nightcrawler capitalists did not intentionally arrange this assemblage. In fact, it is difficult to isolate who or what is actually in control of this industry. The dew worms in the ground, as natural and common as they might seem, are instead products of social, political, economic, and ecological forces. They only exist in their current state (big, fat, and long) in Ontario because of glacial retreat creating clayey loam soils and the colonial ship ballast that dumped the European earthworm species onto North American shores. Now, the most valuable nightcrawlers are found on dairy farms, owned by farmers who operate in a regulatory framework that gives them a guaranteed market and income for their milk. They grow alfalfa as feedstock and spread manure according to provincial regulations. The nightcrawlers blossom under the perennial alfalfa roots and feed on the buffet of cow manure. They grow large, mate, expel nutrients through their digestive tract, and dig tunnels that aerate and filter water through the soil. Understanding why some people pick worms and others do not means that we also have to include not only the Vietnam War, immigration policies, and the specific 1980s economic context in which immigrants arrived in and around Toronto. We also have to include important semiotic elements such as conceptions of leisure time, good soil, good milk, freedom, and autonomy. Critically, none of these individual elements, material or semiotic, have anything to do with the nightcrawler industry. The industry develops and exists only in this set of disparate heterogeneous and autonomous elements, and few of them would change if the industry suddenly vanished.

But what does this mean for our understanding of capital? What does all this co-constitutive, nonhuman labor, work/energy, and unpaid labor mean for the production of surplus value, how it is captured, and how it is accumulated? Is capital merely a parasite living off the life forces of other elements with no internal logic of its own,[53] demoted to one element on equal ontological footing to all the others?

And herein lies the drawback of some uses of assemblage thinking, especially when interrogating the stability of lively capitalist commodities. Relegating capital to one element among others in the assemblage can ignore its role in stabilizing the assemblage. In some discussions around assemblages, mention of capital accumulation remains absent or significantly minimized, or, as Fine notes, incredibly undertheorized (for example, speaking of "capitalization" without reference to "capital").[54] Emphasizing the constitutive dimensions of nonhuman organisms without taking account of processes of capital accumulation risks missing "a vital aspect of their logic and consequences."[55]

By relating the potential benefit of assemblage thinking to the political economy of the nightcrawler commodities, I build on two recent works that have used the assemblage concept as a methodological tool to understand the contingent nature of commodifying biological elements. In *The Mushroom at the End of the World*, Anna Tsing applies the assemblage concept to the matsutake mushroom, a species that has (thus far) defied attempts at commercial culturing. Tsing suggests thinking and writing through assemblages allows researchers to notice the "patterns of unintentional coordination" by "watching the interplay of temporal rhythms and scales in the divergent lifeways." As a method, they become "sites for watching how political economy works."[56] In *Wilted*, Julie Guthman applies assemblage thinking to understanding the political economy of the current conjuncture of California strawberry production. This "explanatory heterodoxy" is important because "not only have the intra-actions of plants, soils, fungi, chemicals, climate, and human bodies shaped the conditions of possibility for strawberry production, but so have tendencies, dynamics, and institutions like profit appropriation, land speculation, regulatory mechanism, and university science."[57]

I use assemblage thinking in similar ways to orient the discussion around how the living organism—with all its idiosyncratic rhythms and processes—is subsumed within the process of capital accumulation. Capital must accumulate surplus value even if it cannot strictly dictate the conditions of its very existence. This is its inner logic. The heterogeneous relations, the temporalities, the contextual daily rhythms, the materialities and ideologies do not always exist because of, nor are they exclusively determined by, capital, but capital must accumulate through them or it ceases to be capital. At an individual level, capitalists have the ability to use, abuse, incorporate, and discard particular elements at particular times, persistently experimenting with socio-ecological arrangements that are conducive to their own perpetuation across a variety of contexts (hence the *variegates of capitalism* literature). But the systemic imperative for the accumulation of surplus value plays a structuring role that subsumes and transforms the generative and reproductive capacities of living organisms, unpaid work/energy, and ecological processes into measurable capitalist value through the labor process.

I use this conjoined subsumption framework through assemblage thinking to empirically investigate and provide answers to the puzzling questions outlined above. Why do people pay for worms? Why this worm, from that land? Why can't *L. terrestris* be cultured like other earthworm species? Why are they picked on dairy farm fields? Why can farmers charge such high rents? Why do nightcrawler wholesalers rent small pieces of dispersed land instead of one sizeable parcel? Why not just buy the land? Why is the industry "underground"? Why must the worms be picked by hand? Why are most laborers Vietnamese? Why do some "actually like" the dirty, repetitive, and exhausting work? Why are workers compensated by the piece? Why is it so challenging to find workers? By seeing worm picking as stuck in the formal subsumption of nature (and all the subsequent implications), we have a clear analytical starting point with the potential to move "beyond the antinomies of posthumanist and political economy inquiry"[58] by tracking how capital accumulates (or not) through the production of "lively commodities." In sum, this account shows how the uneven ability of capital to transform the ecological conditions inherent in the nightcrawler commodity assemblage shapes how investments are made, what soil is necessary, how rent is organized, the working conditions of labor and their payment schemes, and ultimately how surplus is produced and captured.

OUTLINE OF THE BOOK

Very little is known about the nightcrawler industry. "Worm picking" has only been mentioned in the academic literature a handful of times, often as a passing reference as a possible job for recent immigrants.[59] The most detailed account comes from a book chapter that outlines the basic system of production in the late 1970s and early 1980s.[60] There are scant government records, incomplete export values, and no official documents about how the industry began or operates today. As such, most of my research was collected through a qualitative approach; I conducted semi-structured depth interviews with twenty-six farmers, sixteen worm pickers, and eight coolers operating in the industry. I also interviewed eighteen key informants regarding soil dynamics, earthworm ecology, and farm management practices. This included interviews with the Ontario Ministry of Agriculture, Food, and Rural Affairs (OMAFRA), crop advisors, soil researchers, prominent recreational fishermen, bait retailers, and a retired Canadian Revenue Agency agent. Most interviews took place between July 2018 and September 2020, with follow-up interviews conducted between 2020 and 2023.

Several of the interviewees admitted to engaging in illegal activity—largely tax evasion—and as such, I use pseudonyms and provide vague generalizations about their backgrounds only when necessary. Real names are used sporadically throughout only when people have given me explicit permission to do so or when I cite public information—but I refrain from using those same names when they came up in my primary research. I also rely on secondary research collected by OMAFRA for information on farming expenses, crop prices, average yields, and average county land rents. To reconstruct the history of the nightcrawler-as-bait, I primarily rely on newspaper archives spanning over a century (from the late 1880s to 1980s), which covers the period of transition when nightcrawlers went from a self-provisioned bait to a boyhood gig before transforming to a capitalist industry concentrated in southwestern Ontario.

This is where I begin the nightcrawler story. Chapter 1 tells a capitalist commodification story that begins with the cultural politics of recreational fishing—who did it, why they did it, what bait they used, and why. In the nineteenth century, the nightcrawler was considered a "plebeian bait" by fly-fishing aristocratic anglers before becoming the most popular live bait in the post–World War II United States, where a new class of anglers—relatively unbound by the codified practices of aristocratic angling—wanted a bait that was effective, adaptable, and easy to use. In response, southwestern Ontario, with its contingent privileges of loamy clay soil and cheap immigrant labor, quickly supplied the skyrocketing demand. This fortuitous combination of land and labor created an abstract socially necessary labor time for nightcrawler production that could not be matched by other North American competitors.

Once consolidated in Toronto, coolers sought to find an edge over their competition and "followed the worm" to landscapes most conducive to high *L. terrestris* populations. Chapter 2 describes nightcrawler biology and the dialectical relationship between agriculture and worm populations—how each influences the other and how this impacts soil fertility and crop production. As farming practices changed, so too did earthworm dynamics, and nightcrawler capitalists were forced to move from golf courses to dairy farmer fields—the contemporary site of nightcrawler production.

Dairy farmers, however, do not give away their worms for free. Chapter 3 addresses the question of land rent and describes how dairy farms have become enrolled in the production of the nightcrawler commodity, turning the farmers into accidental rentiers. I show how southwestern Ontario farm management practices have unintentionally raised the productivity of labor and the productivity of worms themselves. Farmers have mixed feelings about renting land to worm pickers, as they must determine the ecological value of the earthworm with a monetary value that far exceeds the income from growing other crops. For some, the loss of worms is not worth the risk to their soil; for others, worms are more profitable than growing anything else. Chapter 4 looks at the nightcrawler capitalists—how they came into being and how they coordinate production and manage (or fail to manage) the challenges of making money off of *L. terrestris* physiology and ecology, including the difficulty of fixed capital formation, the geographical dynamics of production sites, low barriers to entry, and control over labor. Based on the materiality of production, nightcrawler capitalists have turned to the underground economy, engaging in money laundering and tax evasion as a means to squeeze out every little bit of surplus value they can to outmaneuver their competitors.

In chapter 5, I go to the worm-picking field to describe the concrete labor practices stuck in the formal subsumption of labor and the contemporary technologies and techniques workers deploy. I also explain why pickers tend to be Vietnamese immigrants, now in their fifties and sixties, why they are paid by the piece, and perhaps most surprisingly, why many pickers feel "free" in the field. Chapter 6 unpacks these expressions of complicated feelings of "freedom" and "autonomy" and how they relate to the ecology of the production process itself. I also discuss the attempt of nightcrawler capitalists to regain control over labor by seeking workers overseas through Canada's Foreign Temporary Worker Program, and whether this will rebalance the power dynamics between pickers and coolers.

Chapter 7 looks at the potential future of the industry and examines how the nightcrawler assemblage is under constant threat of fraying. The very same heterogeneous and autonomous elements that made up the assemblage (the soil, the farms, the migrant labor, the climate, the worm physiology) are all moving in directions that threaten accumulation in the bait industry. In addition to the problem of finding labor, I address the challenges of increasing land rents, climate change, municipal and state regulations, and dwindling market demand.

. . .

No entrepreneur set out to develop a bait worm industry in Ontario; it *emerged* as an industry and took shape through the contingent socio-ecological conditions of production that constituted who worked and who profited. The production of nightcrawler commodities is so interesting because the capitalist's common arsenal of tools—investment in scientific research, fixed capital technologies, breeding techniques, or biological or synthetic inputs to increase yields or neutralize diseases—has not been successfully deployed. In the nightcrawler assemblage, we can see a capitalist industry stripped of the elements seemingly necessary to produce surplus value by directly confronting a stubborn nature. On the surface, this appears as a contradiction. The nonhuman elements of the assemblage appear as the dictators of production, with "capital" merely tacked on as another element in the assemblage, as opposed to a determining force. The contingent socio-ecological elements and processes in the nightcrawler industry necessitate assemblage thinking, and nightcrawlers might seem an ideal commodity to study from a posthumanist perspective.

But demoting capital to one element among many that constitute the assemblage comes with significant analytical risks. Without the pursuit of surplus value, the nightcrawler assemblage would not exist; it would have no meaning or coherence and would instead be a random association of heterogeneous phenomena. Without the imperative to accumulate capital, the Vietnamese immigrant would never cross paths with the Dutch dairy farmers, let alone pick a worm on their land in the middle of the night, nor would I have any reason to discuss any of these disparate elements as constitutive of commodity production. The process of capital accumulation enrolls, adapts, and experiments with heterogeneous elements in the pursuit of surplus value, even when applications of science and technology cannot take hold of and transform the characteristics of nightcrawler ecologies. It shows how capitalist value is produced by and dependent on noncommodified processes, elements, and relationships that cannot be priced through market dynamics. Nature is embedded in capitalist relations and production.

This peculiar hodgepodge shows how the capitalist imperative to accumulate surplus value is the only thing that has been holding these diverse elements together and helps explain how this industry has gone unnoticed for decades while operating under cover of darkness throughout Ontario farmland. By understanding capital's uneven control over nature and labor, we can see why O. R. Jackson's worm trap—an attempt at real subsumption—was doomed to fail, and the demand for more Worm Queens continues to grow.

ONE

From a "Plebeian Bait" to the "Canadian Nightcrawler"

> Fly fishing may be a very pleasant amusement; but angling or
> float fishing I can only compare to a stick and a string, with a
> worm at one end and a fool at the other.
>
> English poet SAMUEL JOHNSON (1859)

THE WORM EXCHANGE

Observant literati may already be aware of worm-picking industry, considering it appeared in one of the most praised novels of the twentieth century. In Michael Ondaatje's award-winning *The English Patient*, two characters are holed up in an Italian hospital at the end of World War II, reminiscing about their past in prewar Toronto. They remember: "The worm-pickers with their old coffee cans strapped to their ankles and the helmet of light shooting down into the grass. All over the city parks. You took me to that place, that café where they sold them. It was like the stock exchange, you said, where the price of worms kept dropping and rising, five cents, ten cents. People were ruined or made fortunes."

Ondaatje reorganizes time to place the Night Crawler Café, as he calls it, prior to World War II. It would not have existed at the time, but like much great fiction, the worm stock exchange is rooted in fact. The Night Crawler Café was actually the Sinistri Café, a small Greek restaurant that served as the epicenter of the worm exchange in the 1970s and 1980s. During these decades, nightcrawlers were picked in pockets across North America, but their price was set in front of this Toronto café.

The worm exchange always began in the early mornings. It had to. The box trucks and Dodge Maxivans had just left the golf courses and pastures fully packed with hundreds of thousands of living, wriggling worms that needed to be sold before rising temperatures would suffocate them. One by one, the trucks parked along the shoulder of Pape Street in front of the Sinistri in

Toronto's Greektown. The odd exchange garnered a bit of attention from a young Ondaatje as well as some curious journalists working for the *Wall Street Journal* who wanted to witness "raw capitalism" at work. Sitting at the café, one journalist observes "50 unshaven men . . . sip thick, sweet demitasses of coffee . . . over games of cards and backgammon . . . and look over their shoulders a lot," keeping an eye out for their competitors, potential sellers, and any tax folks who might be interested their underground activity. They describe the scene as the "worm equivalent of the New York Stock Exchange," attracting all the big wholesalers each morning.[1] As one older cooler, whom I will call Nico, told me, in the 1980s "anyone who is buying goes there."

Clandestine negotiations began at 7:00 a.m. as the first worm trucks trickled in, but no one wanted to be the first one to set the price and potentially overpay for the worms. As more trucks came in and the temperature began to rise, the coy waiting period transformed into urgency: coolers needed worms in their warehouses, and crew chiefs needed to unload the warming commodity. By 10:00 a.m. the pressure was on for coolers like Nico: "It's time to get the fucking worms and get home or get out with no worms."

Coolers zigzagged between crew chiefs and slowly firmed up prices. Another cooler, whom I call Gordon, remembered how quickly the price could rise. Jon might start at $12, then Bob for $13, and another John for $15. Everyone would watch each other, either yelling the prices or keeping an eye on the more secretive coolers, standing next to their trucks in quiet conversations with crew chiefs. As the newspaper article notes, "The same worms may change hands several times in a morning, and the wholesale price may move as much as 20% in a day." It vacillated between quiet negotiations—usually between Greeks, who would break off in their own familial groups—and verbal shouting and "cussing" in broken English largely between the Greeks and Vietnamese. As the worms were sold and moved between trucks, other coolers would have to make rapid calculations, thinking about how many thousand worms needed to be shipped to their customers and how much room they had in their warehouses. Sometimes, with an order to ship the next day, as one cooler told me, they would have no choice: "I'm the stupid one paying $15 per thousand. I pack up 500,000 and go to my warehouse."

Nico remembered one dominant cooler, Ben, who was always suspicious of the Greeks at the exchange, like Nico, who were often his former employees before they broke away and started their own bait businesses. Ben was always animated and was known in the industry as a shrewd businessman, never afraid of conflict. In our interview, Nico stood up and reenacted a conversation he remembered after Ben had given him a dirty look.

NICO: What's going on? Something wrong? Why do you look at me?
BEN: Look how many pills you make me get!

Nico reached into his pockets and pulled out imaginary pill bottles to add emphasis to his story. He carried on impersonating Ben.

BEN: This is for the pressure, this is for the sugar, this is for this, this is for that. You make me keep so many pills in my pocket.
NICO: Why? Why?
BEN: Why? Because you raise the fucking price on worms.
NICO: I raise the price? You raise the price!

After the short drama, Nico sat back down. "In that time, everybody's like that."

By the 1980s, this Greek café had emerged as the nightcrawler epicenter, making Toronto the undisputed worm capital of the world. Millions of worms would be traded each morning at the exchange, ruthlessly establishing the value of the worms in a way that could not be matched by other producers in North America. Nightcrawler capitalists across the continent, from Oregon to Ohio to North Carolina, could not keep up with the flow of worms through Toronto.

Pape Street in Toronto seems a random locality to host the buying and selling of an otherwise ubiquitous earthworm—the same species found in gardens, farmer fields, and sidewalk puddles across the continent. Yet if you enter any convenience store or gas station near a river or a lake, there will likely be a discreet mini-fridge beneath the energy drinks and beef jerky full of dozens and dozens of handpicked nightcrawlers from Ontario—uniformly thick as a pinky finger and ready for the hook. How did this happen?

This is a simple question with a complex answer. At first, the tripartite relationship of fish, worm, and hook appears to us so natural that we forget how unnatural it really is. Earthworms live in the earth and fish live in water; fish need earthworms as much as they do bicycles. Why and when did worms become so popular for freshwater recreational anglers? What material and social transformations were necessary to turn the nightcrawler from a humble worm into a valuable commodity that is picked, sorted, and transported to thousands of retailers across the continent? And why is there only one worm capital of the world?

In this chapter, I tell a capitalist commodification story of *L. terrestris*, how it went from a ridiculed "plebeian bait" to the trusted and reliable "Canadian nightcrawler." This is a historical inquiry that explores how the commodity came to be and how its value is established and maintained by examining the myriad of ways that contingent spatial-temporal heterogeneous elements and processes actively work to ensure successful (and reproducible) commodification. For lively commodities, commodif*ication* inherently raises the question of how nonhuman life forms are shaped by the tendencies of capitalist commodification in pursuit of profit. Anyone in the world can pick up a worm and sell it. Perhaps this is the most basic understanding of commodification—a process that involves taking qualitatively different non-market goods and rendering them commensurable through a monetary price. As simple as this sounds, this presupposes an assortment of historical conditions that can either facilitate or impede processes of commodification.

In the first chapter of the first volume of *Capital*, Marx notes how a commodity is worthless if two conditions cannot be met. First, any commodity sold on the market must have a *use value* that fulfills a societal need or want, "whether they arise, for example, from the stomach, or the imagination." This should be obvious; if no one needs or wants a product or service, there will be no buyers, and the commodity has no value. The sheer diversity of commodities that can be bought and sold in the market is so vast that Marx specifically avoids an extensive accounting of the creation of use values, suggesting such a project is a "work of history" as opposed to the study of political economy.[2] Indeed, the history of why and how people consume the commodities they do was largely ignored by political economists for the first half of the twentieth century, who were more concerned with the macroeconomics of production (though Bourdieu's *Distinction* and Mintz's *Sweetness and Power* are notable exceptions that view production and consumption as dialectically related). The so-called cultural turn in the 1970s, however, reshifted attention to the social meanings and constructions of societal practices, desires, needs, and tastes—often with an implicit (and sometimes explicit) rejection of what was seen as the overly deterministic nature of the laws of capital.[3]

Why, for example, is a diamond or an earthworm valuable? Why in the world would anyone want to put such a rock on their finger, or a worm on their hook, in the first place? The focus of consumption studies has been less on how the commodity was produced and more on the diversity of subjective experiences and social meanings at the point of consumption.[4] The use values of commodities are therefore not ahistorical or natural and exist only at particular historical conjunctures, steeped in mores of what and why certain things are useful. This holds for diamonds as well as worms.

The second condition of a capitalist commodity is that it must be procured through monetary exchange. The worm that is dug up and used by an angler at a riverbank is not a commodity. Nor are the cherry tomatoes I ate from my garden or the haircut I gave myself under COVID lockdown. Such things certainly have a use value, but they only become commodities through a monetary exchange. This monetary exchange gives the commodity an *exchange value*. Under capitalism, commodity production becomes increasingly motivated by this exchange value. Yes, the use value of commodities continues to exist; it must. Otherwise, no one would pay a cent for them. But the capitalist could not care less about the actual use value. What motivates production is that the producer ends the day (or month or year) with more capital than they started with. Production of exchange value becomes an end unto itself.

However, understanding how previously non-marketized living things become sold for a price does little to explain why different commodities have different exchange values. My question is not "what is the price of nightcrawlers in an Indiana vending machine?" but rather "why and how did *L. terrestris* become so thoroughly commodified, whereby effective demand for hundreds of millions of worms developed a competitive market that pits worm enterprises against one another and systematizes production in one particular region of the world?" In this sense, *capitalist* commodification is more than putting a price tag on something that was previously outside of monetary exchange. Capitalist commodity exchange exists when producers are competing with one another in the most cost-efficient way in the same market.[5] This is the enforcer of valuation, where surplus value is generated by persistent efforts to increase productivity and cost-effectiveness through the "coercive laws of competition."[6] Under such conditions, producers are forced to systematize production regimes that increase labor productivity to levels at least equal to industry averages. Failure to do so will, generally speaking, run the capitalist out of business. This is the *law of value* that disciplines producers and generalizes commodity production across an industry.

But what happens when capitalists cannot address the unruly or wild characteristics of a lively commodity, where conscious applications of science, machinery, or genetic manipulation do not increase the productivity of the organism? Does this systemic process of capitalist accumulation cease? If there are no initiators or purposeful investors shaping the ecological conditions of production, does the law of value still hold?

In this chapter, I address these two conditions of commodification—use value and exchange value—to show the historical transformation of the humble dew worm into the Canadian nightcrawler. I track the shifting cultural politics of fishing with worms to understand how the material and social transformation of recreational fishing—who did it, why they did it, and how they did it—shaped the use value of the nightcrawler at specific historical junctures. Between the 1950s and 1980s, the growing demand for nightcrawler bait forced suppliers around North America into fierce competition. However, the inability to control the ecological conditions of nightcrawler production relegated the most productive producers to southwestern Ontario, where the contingent privileges of useful soils combined with cheap and efficient labor. This explains how and why the sale of nightcrawlers went from an impossibility in the nineteenth century to a boyhood gig in the early 1900s before developing into a peculiar capitalist production system by the mid-1950s and consolidating in a morning auction held in Greektown in Toronto. Seeing how nightcrawlers went from not being a commodity and having no capitalist value to becoming a capitalist commodity whose value is determined by socially necessary labor time displays how contingent, open-ended, and unplanned the process of commodification can be. However resistant and unruly the worms, soils, and laborers might be, nightcrawler capitalists across the continent competed against one another and established a socially necessary average labor time that could only be maintained in southwestern Ontario.

GENTLEMAN ANGLERS

As synonymous as the worm and the hook are today, these two conditions of commodification were not met in the early days of recreational fishing. It is not entirely clear when fishing for food became fishing for fun, as the two objectives are not mutually exclusive. Archeological evidence of fishing for food predates our own speciation. There is evidence of fishing in Africa from over two million years ago, likely consisting of grabbing the catfish that gathered along shallow pools. Neanderthals also appear to have been skillful spearfishermen. We humans perfected the art of "angling"—understood as the process of fishing with a rod, line, and hook—with the advent of the hook at least 45,000 years ago.[7] Drawings of Egyptian high priests suggest they were "having fun" fishing for tilapia in large stone tanks,[8] and there is a reference to Emperor Louis the Pious, who went on several fishing excursions in the ninth century—most likely not to fulfill his caloric necessities.[9] The earliest explicit references to fishing for the joy of it come from twelfth- and thirteenth-century sources that suggest commoners took pleasure in fishing with a hook.[10] In fact, in these sources, it is often the "commoners" that must explain the practice of angling to the upper classes. These early written sources "independently portray a pastime of non-aristocratic origin but hopeful pretensions, for each seeks to promote the participation of the social elite."[11]

Such "hopeful pretensions" quickly succeeded, and by the sixteenth century, recreational fishing was clearly an aristocratic affair. The mysterious author of the famous 1496 *Treatyse of Fysshynge Wyth an Angle* makes clear fishing with a line was not for the purpose of calories but as a means to "enduce a man in to a mery spyryte." The instructions in the *Treatyse* were so detailed and clear that its publisher worried it might attract too much attention from the "commoners." Instead of publishing the pamphlet for mass appeal, he appended the document to a larger volume on hunting that targeted the more "gentyll & nobel men," ensuring those of "lytlyll mesure" could not "vtterly dystroye" the joys of such sport.[12]

Although more affordable than hawking or deer hunting, recreational fishing still required access to lakes, streams, and ponds—access which, at the time in England, was quickly disappearing. The Enclosure Acts further solidified the social distinction reserved for them by denying commoners access to rivers and lakes, and a market system developed to access them.

The increasingly private ownership of riparian rights greatly shaped the social composition of English fishermen: "Once established in private hands, fishing rights evolved, mainly through rentals and leases, toward a market commodity," with lords managing their private fisheries for rental.[13] Recreational fishermen had the option to pay for the right to fish, or not fish at all, with landowners on a good stretch of river charging exorbitant fees for the right to catch the "game fish" like trout and salmon.[14]

In early sixteenth- and seventeenth-century publications, worms were considered a suitable bait—alongside grubs, crickets, and artificial flies—if used appropriately in the right season and for the right species of fish. In the eighteenth and nineteenth centuries, however, recreational fishing practices began to change. The privatization and exclusivity of recreational fishing "rapidly accumulated a burden of tradition and convention," specifically around the types of baits and lures.[15] Fly-fishing in particular added a costly artistic flair to an increasingly exclusionary pasttime. Such convention and tradition of recreational fishing further entrenched a class divide, not only between those who trespassed on private property to fish for food and those who paid to fish for fun, but between those who fished with live baits such as grubs, crickets and worms and those who fished with hand-tied, intricate, artificial flies constructed from recipes requiring pricy feathers from exotic rare birds.[16]

In the mid-nineteenth century, an angler might have been permitted to use "coarse baits" such as worms or crickets if the conditions were not conducive to fly-fishing. However, by the end of the century, thanks to people like Frederic Halford and the prestigious Houghton Fishing Club, only artificial flies were permitted as bait and were to be used only through a set of "intricate rules."[17] Halford called baiting a hook with a worm "unsportsmanlike" conduct for an angler, only slightly less reprehensible than using a large net.[18] A hierarchy of fish developed, with the "stately" trout and salmon on top, considered "sport fish," and the lowly carp and catfish at the bottom, considered "coarse fish."[19] As recreational fishing clubs increased alongside a burgeoning bourgeoisie in the 1800s, fishing techniques became codified.[20] Fishing with the artificial fly became sacrosanct; fishing with worms, profane. As a writer in the English outdoor magazine *The Field* proclaimed, "In England, we think it is poaching to fish for trout with worms, and I know several men who would not speak to you again did they know you had used that bait."[21]

The codified ideas of fly-fishing that were born in England were quickly embraced across the Atlantic, where the absence of enclosure laws and the usurpation of indigenous lands opened the rivers, streams, and lakes to classes of anglers that would not have had the financial resources to fish for the stately trout or salmon back in England. Elitist ideas about bait transferred across the Atlantic; wise and knowledgeable anglers were fly fishermen and only the "meat fishermen" used worms. As early as 1832, writers in North America lauded the gentility of fishing and lambasted the worm anglers for their "unsportsmanlike characteristics . . . [and] practices subversive of all the chivalric spirit which should animate true sportsmen."[22] In 1891, a commentator in the *New York Times* stated unequivocally, "Bait fishermen are not real anglers."[23] Norman McLean's classic story "A River Runs Through It" (and the popular Brad Pitt film) is emblematic of this mindset. Set in Montana around World War I, the story indicts worm-fishermen and all they represent—poseur recreationists detached from their roots who have neither the skill, spirit, nor patience to appreciate the art of fly-fishing. The protagonist of the story scoffs at a visiting brother-in-law: "I won't fish with him. He comes from the West Coast and he fishes with worms."[24]

Worms certainly had a use value for attracting fish, but they had no exchange value. Flies were what respectable gentlemen (and presidents) cast; worms, on the other hand, were for the crude and unsportsmanlike commoners who could simply dig up this lowly "plebeian bait" along the riverbank.[25] After World War II, however, everything changed.

. . .

Prior to postwar industrialization, there was little demand for the worm as a commodity, nor was there a wide consumer base with the money to pay for it. After World War II, however, recreational fishing boomed in the United States, where industrialized work regimes, a Fordist pay scale, and an infrastructural network increased access to lakes and rivers. The "weekend," in particular, as noted in De Grazia's classic book *Of Time, Work, and Leisure*, created a "dash for the outdoors," as the new working class took to outdoor leisure activities.[26] Recreational fishing in particular caught the attention of the US Fish and Wildlife Service, which conducted the first *National Survey of Fishing and Hunting* in 1955. At that time, the survey concluded, there were 18 million recreational freshwater fishermen, spending over a billion dollars annually on fishing trips, equipment, bait, and fees.[27] These figures steadily increased with every five-year survey, culminating at 39 million freshwater anglers in the United States by 1985, the era when Nico and Ben fought over worms at the Sinistri Café.[28] Indeed, this new rank of amateur fisherman wanted a bait that was effective and could be easily handled. Unburdened by the codified conventions of fly-fishing, demand soared for the so-called "plebeian bait" that was adaptable to the most diverse of fishing circumstances. The nightcrawler fit the bill. "When nothing else works," experts said (as recorded in *Popular Mechanics* in 1962), "try a worm."[29]

Something had changed in the middle of the twentieth century. Recreational angling was no longer reserved for the propertied lord of the manor, the new moneyed bourgeoisie, nor the fly-fishing purists; it was for the masses. Fly-fishers—with their exclusive tradition and convention that had specifically been kept from the so-called commoners—scoffed at these new fishermen who could not be expected to learn mastery over technique and ecological sensibilities. Surely, the masses of lumbering factory workers with sporadic long weekends could not be expected to enter the fly-fishing ranks and painstakingly learn the wonder of their ways. And the masses did not. When they went out to the lake on the weekends, they fished with the unrefined skill of children; they fished with worms.

Fly-fishers suggested these "live bait" fishermen, with their worms, minnows, and grasshoppers, were vulgarizing their stately recreation. But there is no doubt that worms were an effective bait. Calvin Coolidge was ridiculed by his Republican opponents for using "garden worms" to catch the stately trout at a local fishing derby. In fact, Coolidge caught twice as many fish and "triple the weight" as his fly-fishing competitors.[30] The *New York Times* also tells the story of Albert, a fly-fishing "purist" at heart, who, after having some trouble catching trout, placed a worm on his hook. Though he caught four trout, he could not "meet the gaze of his fellow fly fishermen" from the shame of using worms. The next day, Albert promised himself, he would fish nothing but flies.[31] The former *New York Times* outdoor columnist Nelson Bryant observed a similar ideology when he saw out-of-luck fly fishermen sticking with their dry flies without any bites. Initially, he thought, "Their rigidness led me to believe that they were members of the cult that believes it reprehensible to use anything but a dry fly." Later in the evening he realized he was wrong: "At dusk, the two anglers paddled their canoe to a huge boulder on the shore, lit a lantern, affixed worm-baited hooks to their fly-lines and began drinking beer."[32] Even the fly-fishing Herbert Hoover confessed, "toward the end of the day, when there are no strikes, each social level collapses in turn down the scale until it gets some fish for supper."[33]

Does this mean that dew worms are the best bait for angling? Not really. A quick scan of the outdoor columnists, fishing guides, and online fishing forums will quickly show there is no such thing as the "best bait." What bait is best depends on all sorts of factors: the season, the body of water, the current, the depth, the time, the fish species, and so on. It would be impossible to say any one kind of bait is best. But what if I were to ask the question differently? If a recreational angler could choose to bring only one bait— artificial or live—and travel to an unknown waterway to catch an unknown species of fish, which one bait would they bring? Which bait is most universally adaptable and able to attract the most species of fish in the most diverse of environments? In an 1867 letter to the editor of *Turf Field and Farm*, a New York–based journal, a Mr. Buxton writes, "I do not mean to say that the worm will invariably beat minnow and fly, but it is always a great rallying point and something to fall back upon."[34] There may be no such thing as the "best bait," but the worm is always there, ready to step in whenever called upon. The simplest answer is again proposed by *Popular Mechanics*: it "will catch fish when nothing else works."[35]

But why the Canadian nightcrawler? It is not the only bait worm around. Its closest annelid competitor has traditionally been the red wiggler, or compost worm (*Eisenia fetida*). While it is cheaper than the nightcrawler and can be commercially cultured, fishermen are not as convinced of its use value. The smaller size of red wigglers tends to excite only the smaller fish. They are more difficult to get on the hook and die faster in the water. They are too "thin, brittle, and difficult to keep on a hook without damaging them" and anglers go through more of them in a short time.[36] Handling the red wigglers will also induce the secretion of pungent mucus which, in contrast to the nightcrawler, can often repel fish. While there is no doubt the red wiggler can work under the right conditions, fishermen are less enthusiastic, merely willing to "settle for the skinny worm."[37]

The Canadian nightcrawler's use value stems from its specific physical characteristics: it is big, wriggly, aromatic, and delicious. It is the longest and thickest of the bait worms, which increases visibility and attracts the larger fish. Once on the hook, it wriggles mightily and maintains this continuous movement long after it has been cast or dropped into the water.[38] Nightcrawlers also have an attractive scent for fish. They secrete a calciferous mucus from their skin membrane when stressed (i.e., impaled on a hook), which anglers suggest the fish can recognize. And the taste cannot be easily replicated.[39] Many fishermen try to make useful analogies for their non-fishing audiences: "It's like a pizza for worms" or a "filet mignon."[40] Artificial worms—first patented in 1951—have slowly and successfully mimicked these attributes, even adding new designs, scents, and tastes. But as one accomplished and professional fisherman said to me, "Why would you imitate live bait when you can *have* the live bait?" He went on: "Even though I fish with artificial lures, I've always thought that 95 percent of fishing, if you can properly present a live bait, it will outproduce anything else."

Post–World War II recreational fishermen, unbound by fly-fishing conventions, wanted a bait that worked. The nightcrawler's physiological characteristics—its length, its wiggle, its girth, its scent, and its ability to catch the widest array of fish—generated a specific use value for tens of millions of recreational fishermen who had a free weekend and some cash to pay for it. The nightcrawler commodity became possible.

TROUBLESOME BIOLOGIES, ECOLOGIES, AND PHYSIOLOGIES

Wayne King converted an old gas station into a worm warehouse that would keep the nightcrawlers cool in the otherwise sweltering North Carolina weather. "Here in Hayward Country," he told a journalist in 1988, "is the only place in the South that I know of where there are worms produced."[41] He employed hundreds of worm pickers and soon became known as North Carolina's "earthworm king," producing 50 million nightcrawlers a year at the height of his business. "Thing is," King said, "we're getting smaller and smaller every year." By 1988, he was down to 30 million nightcrawlers, which was not enough to supply the demand from his customers. He covered the shortfall by importing nightcrawlers from Canada. He was well aware of the industry in Toronto and knew he could not keep up, noting, "There's more worms that go into Toronto after a good night's picking than all the worms produced in North Carolina in a whole year."[42]

King found himself in a losing battle against the Torontonians who were persistently outpicking their rivals across the United States. Attempts at industrially farming the nightcrawler failed. So did the traps, liquids, and electroshock methods that sought to increase labor productivity. As a result, the most efficient way to "produce" a nightcrawler remains the most rudimentary: bend over and pick it up.

For early US bait fishers, this was part of its appeal. The worms could be secured by anyone willing to dirty their hands. In the nineteenth century, fishermen would often bring a shovel to the shoreline and dig for their own supply: "It has been estimated that during the past twenty years, there has been enough dirt removed and labor expended in the quest of fish bait to have completed the Panama Canal."[43] The fishermen would strike the shovel into the dirt, overturn the earth, and pick through soil and debris to find the suitable worms. It was not easy work, with many fishermen having "lamed their backs," blistered their hands, and "wasted their time."[44] If you know fishing, one newspaper said, you know "how a fisherman hates to dig bait."[45]

There have always been other ways to get the worms to the surface. One suggestion was to put a stick in the ground and grind it like a washboard, which would "set the worms in motion." Other fishermen found pouring certain substances over the ground would drive the worms up and save time, blisters, and backaches. This included soapsuds, cupric sulfate, and, more recently, xylene. Earthworm researchers currently use a mustard powder solution to bring worms up to the surface. The problem with each of these methods for harvesting nightcrawlers is that all worm species might come up; pouring liquids thus adds the step of sorting the useful bait from the useless.

Nightcrawlers, however, are called nightcrawlers for the simple reason that they come up at night of their own accord. Digging or pouring substances is unnecessary if one adheres to nightcrawler behavior. In 1891, some New Haven, Connecticut, residents believed the city's new electric lights acted "as a magnet," as they witnessed thousands of worms crawling throughout a city park.[46] More empirically minded folk, however, suggested the lights only made visible the natural behavior of these worms; the nightcrawler comes up all by itself. In 1896, an article described a nightcrawler-picking process that has remained largely unchanged for over a century: "Some men go around with a lantern on favorable evenings—warm and slightly rainy weather is best—and find the worms crawling on the worn or barren places."[47] Other newspapers provided detailed instructions on how to procure this bait. The "easy method of gathering them is at night, when they are above ground, especially after there has been a warm shower. You will need a lantern and a pail. Step along softly, holding the lantern near the ground, and hundreds of worms will be disclosed in the wet grass."[48]

Not every fisherman wanted to waste their nights collecting bait, and many preferred to pay someone else to do it for them. They did not have the "inclination to dig up the back yards in search of the worms or creep about on hands and knees looking for night crawlers."[49] Kalamazoo businessmen, the *Grand Rapids Press* wrote, simply "have no time" to pick worms.[50] It was also a dirty labor process that was "rather undignified for a businessman." Instead, they paid others to procure the worms; this tended to be "boys." In 1912, the *New York Times* described "fish bait" as "one of the scarcest commodities in the city," with many "small boys working overtime digging worms."[51] In New Haven, some "citizens have noticed with a great deal of curiosity, groups of boys at work in the nights" collecting the useful bait for fishermen.[52] In 1896, a newspaper article notes with surprise how "a small boy will get a can-full [of worms] for a quarter."[53] "Angle worms are a commodity and 25 cents a can the price," says another article, noting "the angler hails the bait seller as a friend."[54]

However undignified the labor, selling worms appeared surprisingly profitable. "The custom of baiting hooks with worms when angling for fish runneth back to time woeth not," starts one journalist, "but the custom of using worms for bait when fishing for capital is believed to be entirely original with these youngsters."[55] At railway stations, "scores of fishermen leave the city every day for nearby lakes and ingenious small boys have discovered that good bait has a ready sale."[56] "A keen competition has sprung up at some of the railway stations between rival dealers, but it is not so strong that there is any reduction in price, for 25 cents is always asked for a can."[57] A few years later, however, the price dipped: "Worms! Worms! Only a dime a can!"[58] In 1919, a sixteen-year-old picking and selling nightcrawlers made $65.60 US in one day in this "new and unique business." The *Grand Rapids Press* notes for the first time that some people have begun making a living picking worms, generating between $7 and $20 US per day.[59] In the same article, there is the first mention of a rudimentary network of worm pickers and dealers that transport the worms across the country.

In these early articles, the worm pickers appear to be their own bosses and lack clear capitalist social relations with a value determined by socially necessary labor time. The "youngsters" and "boys" picked worms and sold them in their own regions, essentially fragmenting the nightcrawler market. There are several articles, however, that suggest the profit incentive began to shape social relations on increasingly capitalist terms as worm pickers began to work for a wage, with some of the surplus going to their employers instead of themselves. In 1926, Charles L. Carroll of Springfield, Massachusetts, became the self-proclaimed "worm king," selling over 20,000 nightcrawlers in a season. He employed a "regular force of boys and young men, equipped with pails, slickers and carbon lamps," who would search the lawns and fields for nightcrawlers.[60] The *New York Times* describes another capitalist who employed ten "youngsters" to "carry on the production end of the business."[61]

Records of these early worm sellers and proto-nightcrawler capitalists are scattered across the continent in the early twentieth century. While any "youngster" could make a few dollars selling in segmented markets, demand for the bait slowly put these budding nightcrawler capitalists into competition with one another. Mid-century newspaper articles show that worm wholesalers sprouted up across North America, where suitable soil types and appropriate climatic conditions could support dense nightcrawler populations. In Canada, there is a report of a nascent worm picking industry in Ottawa in 1951 that describes a worm wholesaler, Pete Therein, who decided to close the city's "biggest live bait business" when his four worm pickers demanded higher wages. Classified ads show a need for worm pickers around

Ottawa and as far north as Montreal, especially in the early 1960s. In Vancouver in 1957, worm picking occurred along the boulevards and in public parks, with classified ads seeking "experienced dew worm pickers with own transport."[62] Such ads ran throughout the 1960s to 1973, paused, then appeared sporadically between 1983 and 1987.

Pockets of picking developed across the United States as well. In 1950s Oregon, there is talk of "nightcrawler hunters" getting ready for the start of fishing season,[63] and other complaints about pickers trampling the city of Salem's rose bushes.[64] Classified ads in the state ran throughout the late 1950s into the 1980s, and three relatively large US nightcrawler companies were established.[65] In Idaho, the *State Journal* ran ads seeking pickers beginning in 1977 and mentions a few larger businesses capable of selling over a million worms.[66] And as we already know, in North Carolina, the industry had consolidated in the hands of Wayne King. There are also classified ads and a few smaller companies in Wyoming and Utah, but the majority of US picking occurred around the Great Lake states. In the 1940s and 1950s, there were ads for worm pickers in Chicago, Port Huron, and Rochester. In 1950s Detroit, greenskeepers began complaining about worm pickers on their golf courses at night,[67] a common irritant that would continue for decades.[68] Illinois classified ads run from 1974 to 1987. In Indiana, one ad states "highest price paid by me." In Ohio, an employer in 1954 was looking for "experienced worm pickers,"[69] with such ads running until 1978.

Wholesalers in the region were confident their soils were good habitats for the nightcrawlers, but they always ran into a problem: "Our people aren't motivated for that kind of labor."[70] The description of the pickers in these articles would not instill much confidence in a business owner aiming to hire the most productive laborers. As Mrs. Welty, an Oregon producer, stated in 1970, "We got enough night crawlers here in the Willamette Valley for the whole world—if we could only get enough pickers." Her four hundred part-time pickers are described as "retirees, college students, and enterprising children."[71] But, she says, "elderly women do it best."[72] Other companies describe their labor force as "college kids," "youngsters," "students, housewives, moonlighters, and senior citizens."[73] In the mid-1990s, one producer said that more than half of his twenty-five worm pickers were between the ages of ten and eighteen.[74] Some places clearly had enough worms, but they lacked productive pickers, while others may have had productive pickers but lacked worms. In one place, these two conditions profitably collided.

Dave and John Brennen, southern Ontario brothers, started picking night-crawlers in 1937. Like many other pickers across the continent, they picked worms from lawns, boulevards, and parks to sell to local vendors. When demand increased after World War II, they began to formalize their produc-tion and distribution through a network of pickers, trucks, and refrigerated warehouses. By1955, the Brennan Brothers Bait Co. was the largest supplier in the world, exporting 15 million nightcrawlers to the United States, with the *Globe and Mail* noting, "Worms are becoming a big business."[75]

How did the Brennan brothers quickly rise to the top of the worm heap? What did southwestern Ontario have that other nightcrawler capitalists around North America did not? In short, fertile soil and cheap, productive labor. The capitalist nightcrawler industry emerged in southwestern Ontario not through intentional capital investments but through contingent and provisional elements that happened to profitably align at the optimal time. No amount of investment or entrepreneurial creativity outside the region could compete with Toronto's combination of productive soil and labor. Worm producers in North Carolina, Oregon, and Ohio were simply out of luck.

·   ·   ·

Southwestern Ontario soils are specifically classified as gray-brown luvisols,[76] a type of fertile soil rare in Canada. In other soil classifications, they are more broadly defined as orthic and albic luvisols,[77] similar soils to those that sur-round the Great Lakes, created by glacial melt approximately 12,000 years ago. Over the centuries, available organic matter, the loamy clay texture, and temperature, acidity, and moisture have allowed *L. terrestris* to thrive in the region.[78] But these soil types are not dissimilar to pockets in the United States (particularly the Great Lake states), and one would expect to observe similar *L. terrestris* populations. In the 1970s, Alan Tomlin, a former agricul-ture scientist with the Canadian government, suggested the first nightcrawler wholesalers were Americans seeking to fill the demand from recreational walleye anglers around Lake Erie. As recreational fishing greatly increased throughout the 1950s and 1960s, nightcrawler producers moved into Canada looking for more worms and cheaper labor. They found both.

Southwestern Ontario typically features a thicker snow cover over a longer period in winter compared to neighboring US states. Snow cover provides an insulating sheet and allows nightcrawlers to remain active and contribute to soil processes throughout the winter while also protecting the earthworms from predation.[79] Moisture content is also important for *L. terrestris* biomass and density,[80] and the melting spring snow may provide a slow-release "annual soaking" that protects against possible spring droughts. Slightly higher summertime temperatures in the United States also affect the picking nights. Coolers are adamant that their pickers never pick worms when temperatures rise above 17°C (63°F), as the worms can become lethargic, overheated, and prone to morbidity. Such climatic differences possibly allow the nightcrawlers to live a bit longer, grow a bit larger, and reproduce a bit more easily in their Ontario habitats.[81]

As the Oregonian nightcrawler capitalist noted above, certain soils could possess all the nightcrawlers in the world, but it is all for nothing without able-bodied productive laborers willing to pick at a competitive price. Where did the Ontario nightcrawler industry get all of its labor? Between 1956 and 1976 Canada took in close to one-third of all immigrants into North America, despite being one-tenth the size of the United States,[82] with many ending up in and around Toronto. The 1952 Immigration Act sought out immigrant labor to supply the burgeoning urban-industrial workforce. In Toronto, massive infrastructure and housing construction projects proliferated, and the city began taking more immigrants from outside Great Britain and accepting people from southern Europe.[83] Canada's Italian-born population, for example, grew threefold in the 1950s, with over half residing in Toronto. The 1956 Hungarian revolution also produced a wave of immigration into the region.[84] Generally, over half of the incoming immigrants into Canada headed to Toronto; by 1957, one in five people in the city were "new Canadians."[85] In the 1960s and 1970s, "other immigrant groups followed suit. Even as Italian immigration continued, Greeks, Portuguese, and the peoples of the Balkan Peninsula began arriving in Toronto in large numbers."[86]

A demographic history of worm pickers between the 1950s and the 1980s largely mirrors the cycle of immigrants who settled in the Toronto region. Italian and Hungarian pickers were replaced by Greeks and Portuguese, who were subsequently replaced by Southeast Asian immigrants beginning in the early 1980s. Each cycle of new immigrants to the Toronto region quickly took to worm picking, as it provided an opportunity to make significant amounts of money with few language abilities or special skills. Nico remembered flying into Toronto's airport as a new immigrant. His family picked him up and drove directly to a golf course to pick worms. After a brief sleep, he started his first day in Canada with $200 in his pocket.

From the 1950s to the 1980s, immigrants mostly picked on golf courses. The short-cut manicured grasses increased the visibility of the worms, and the constant water supply ensured worms would come up even in the driest of weather. Bait companies would pay the golf courses between $1,000 and $5,000 for exclusive picking rights, which were not insignificant sums of money at the time. Not only did the added revenue help cover maintenance costs for the golf courses, but picking, it was thought, helped reduce the number of worm castings or "middens" that would interfere with rolling golf balls and muddy the greenskeeper's mowers. By 1957, the Brennan Brothers claimed picking rights on fourteen golf courses in Toronto. They increased their warehouse capacity to 15 million worms, selling over 40 million a year. Their goal, they told a reporter in 1956, was to sell 100 million worms to the United States.[87] The piece rate for the pickers at the time tended to hover around $3.50 per 1,000 worms (approximately $36 in 2023 dollars, adjusted for inflation) with some of the better pickers making $30 per night during productive nights (over $300 in 2023).[88]

On an average night, pickers could collect between 3,000 and 5,000 worms.[89] However, on optimal nights, when the temperature and humidity were suitable, and when there was a lack of wind, the worms would come to the surface en masse, and fast pickers could grab 16,000.[90] In 1956, Dominion Live Bait said they employed over 200 "mostly immigrant" pickers and sold 40 million worms to the United States.[91] By 1961, the Brennan Brothers had rented twenty-seven golf courses and were shipping over 60 million worms across the border.[92] In the mid-1960s, Bob Conroy, another early worm baron, claimed he personally made $35,000 a year selling 80 to 90 million worms a year (approximately $290,000, adjusted for inflation), employing 300 full-time and 400 part-time staff who were almost entirely Greek and Portuguese immigrants.[93] Joseph Haupert, an Austrian immigrant who came to Canada in 1961, started a worm-picking business after seeing pickers hunting for worms around his neighborhood. In 1966, he signed a lease with a golf course, and by 1968, he had 435 pickers, including Canadian-born residents, as well as Portuguese and Greek immigrants.[94]

In the 1960s, bait companies reported that an "average" picker could take home between $100 and $125 per week during the April-to-October picking season (approximately $850 to $1,000 today). Better pickers were making over $200 per week.[95] In a 1964 article, the journalist described one Portuguese woman as "the Mickey Mantle of the worm-picking league" because she picked 50,000 worms in one week, taking home $250 (over $2,300 in 2023).[96] With the price of a dozen worms in the 1960s at $0.25 to $0.35 US, pickers were fully aware of their value to the wholesalers and would threaten to work for another company if piece rates were not increased. There were several informal strikes in the mid-1960s, with pickers demanding a $0.50 piece-rate wage increase.[97]

The productivity of pickers rose further in the mid-1960s and the early 1970s. Dave Brennan told a reporter that good pickers "with experience" could get 12,000 worms a night.[98] Others said the average picker on a good night could readily pick 10,000 worms.[99] In 1972, Bob Drouin paid a man for picking 22,500 in a night, "the so-called record" at the time.[100]

Other regions could not compete with the productivity of Toronto pickers. Wholesalers in the Great Lakes states said their pickers might get between 4,000 and 5,000 worms on a good night in the 1980s.[101] These paltry numbers plummeted further as picking spread out. In the '70s, for instance, Idaho and Oregon pickers could grab only a measly 2,500 to 3,000 *per week*.[102] The nightcrawler capitalists outside of Ontario knew they could not keep up. By 1983, an Idaho producer suggested at least 90 percent of US-sold nightcrawlers were coming from Canada. Oregon-based Rainbow Northwest Worms said the same thing: "Canada gets them picked so cheap we can't compete."[103]

. . .

Tracking the production of *L. terrestris* value provides theoretical clarity on the process of capitalist commodification of a lively commodity whose ecological conditions of production inhibit control over nature. There is no single pathway toward the commodification of living things. It is a process contingent on the spatial and temporal alignment of social and ecological processes, both material and discursive, controllable and uncontrollable, that allows capital accumulation. In short, southwestern Ontario provided the land and labor necessary to turn this side hustle into a valuable business regulated by the capitalist value of the commodity. Bait-and-tackle shops and anglers around the continent quickly realized the biggest, juiciest, most affordable worms came from Ontario, and *L. terrestris* was rebranded as the

"Canadian nightcrawler," a commodity whose value was set by the morning worm exchanges in Greektown, Toronto.

But the capitalist commodification story does not end there; it cannot. Toronto had become the worm capital of the world, but intraregional competition forced nightcrawler capitalists to find ways to increase productivity further. To do so, however, they had to follow worms to whatever landscapes permitted them to grow the fastest and, by necessity, learn more about golf course herbicides, pasturing cattle, and the changing dynamics of the Canadian dairy industry.

TWO

Following the Worm

We've never contacted worm pickers. I would say it would have
been in the mid-'80s when they first started coming around.
They were just saying . . . "Can I pick worms in your pasture?"

CHARLIE, dairy farmer

CHARLES DARWIN MIGHT HAVE DISCOVERED the origin of all spe-
cies, but his final work was about only one: the nightcrawler, a lifelong passion. Darwin's book *The Formation of Vegetable Mould Through the Action of
Worms, with Observations on their Habit* (1892) reads like the journal of an
old man who finally had enough time to explore a persistent curiosity. He
wrote to a friend, "the subject to me has been a hobby horse, and I have perhaps treated it in foolish detail."[1] His experiments, it should be said, were no
more complex than those at the sixth-grade science fair. He tested their hearing with "shrill notes" from a whistle and the deep notes of a bassoon. He
found the worms were "indifferent to shouts" and made no reaction to a
piano played as loud as possible. Placing the worm-filled pots on top of the
piano, however, he found the worms retreated into their burrows at different
rates depending on the position of the notes played. He tested their sense of
smell: "They exhibited the same indifference to my breath whilst I chewed
some tobacco. . . ." He shone a light on the worms and noted: "Their sexual
passion is strong enough to overcome for a time their dread of light"—an
observation that I have made myself (wait for it in chapter 6). He watched in
amazement how their castings stimulated the proliferation of weeds around
his flagstone pathway, so much so that his gardener eventually gave up on
landscaping and allowed the path to be swallowed up by the vegetation. He
calculated that 50,000 worms could produce 18 tons of castings per year,
leading him to conclude, "without this humble creature, agriculture as we
know it would be difficult, if not wholly impossible." *The Formation of
Vegetable Mould* would become Darwin's most popular book, and his
research "helped open the door for the modern view of soil as the skin of the

earth."[2] Only in the past few decades have scientists focused on the intricacies of the life-forms that exist inches below our feet, and Darwin's work has been going through a rediscovery among pedologists studying the complex relations between *L. terrestris*, soils, and agriculture.[3]

Darwin's observations about *L. terrestris* might have been profound at the time, but they provide no knowledge that a worm picker or nightcrawler capitalist wouldn't already possess. In their struggle to increase productivity, the nightcrawler capitalists—whether they know it or not—have inadvertently provided new information about nightcrawler biology, physiology, habitable landscapes, and rates of reproduction. In the 1970s and 1980s, with the worm stock exchange in full swing, golf courses were not supplying enough worms to meet demand. Coolers needed to expand to increase profit. Unlike the limited number of golf courses in the area, pastureland appeared unlimited in southwestern Ontario. The resource-rich and undisturbed soils provided a suitable habitat for sustaining high *L. terrestris* populations. Production could be extended beyond the golf courses, and the supply of bait worms increased. However, accessing these fields forced the coolers and crew chiefs to engage with a new actor now controlling the ecological conditions of production: the dairy farmer.

In this chapter, I explain more about nightcrawler biology, why the worm is a recalcitrant lively commodity, and how the relationship between agricultural management practices and nightcrawler population dynamics enrolled pastureland into the nightcrawler assemblage. I begin by describing and comparing the physiological and ecological characteristics of the nightcrawler with other bait worm species to see why it has been so troublesome to commercially cultivate. I then describe how *L. terrestris* unintentionally established itself in North American soils through processes of colonization and how its invasive attributes, though detrimental in some forest ecologies, provide numerous ecological services for temperate zone farmers, who are often intent on maintaining or increasing their nightcrawler populations. This dialectical relationship between farmers and *L. terrestris*—a give and take of practices and services—helps explain how and why, specifically in the 1970s and 1980s, nightcrawler capitalists could find adequate *L. terrestris* populations in southwestern Ontario pasturelands. I then describe the informal (and often conflictual) relationship that initially developed between the dairy farmers and the worm pickers. I conclude the chapter by noting how a significant transformation in the dairy industry—converting pasturelands to alfalfa fields and feeding cattle in barns—had more unintended consequences for the ecological conditions of nightcrawler production, which in turn reshaped the relationships between the pickers and farmers once again.

By following the worms from colonial ships to alfalfa fields, we can see how the conditions of nightcrawler production are not easily controlled by the nightcrawler capitalists. Instead, they are dependent on dynamics and transformations in other industries that have little to do with the bait trade itself. The production process has historically relied on conditions not of its own choosing and has been maintained through shifting assemblages. They must persistently follow the nightcrawlers to environments conducive to higher densities, enrolling new autonomous actors and landscapes that, just like the earthworms, cannot be easily controlled.

WHY NOT GROW *L. TERRESTRIS*?

Hugh Carter, Jr., was a campaign advisor for his cousin Jimmy Carter's early political career. When Jimmy left the state senate, Hugh successfully ran for the same district and held the seat between 1967 and 1981. Both men were also farmers. With his rise to the presidency, Jimmy became the most famous peanut farmer in history. Hugh is perhaps the most famous worm farmer. He began farming red wiggler worms (*Eisenia fetida*) in 1949 in Plains, Georgia, and by the mid-1960s, he was the self-proclaimed "world's largest worm farmer," stocking upwards of 250 million worms and selling between 20 and 30 million each year to recreational anglers. He quickly captured the recreational-fishing bait worm market and maintained his market share over the decades through shrewd business acumen. But competition in the red wiggler bait market was fierce, and others were able to produce millions of red wigglers of their own, all aiming for the same burgeoning market of anglers who looked on these worms as a good bait, although still second-class compared to the bigger nightcrawler. Why didn't Carter use his available capital to shift his production towards nightcrawlers? If nightcrawlers are better bait, why not supply the market with the most useful, popular, and valuable bait?

I spoke with Jim Shaw from Uncle Jim's Worm Farm—currently one the largest red wiggler sellers on the continent—about his own early attempts at culturing the nightcrawler. As a teenager in the 1970s, he was making handsome profits personally picking the valuable nightcrawler from Pennsylvania lawns and mixing them in with red wigglers purchased from Hugh Carter—

whom I thought he called "ordinary." "Ordinary?" I asked. "Ornery!" Jim clarified with force. "A pain in the butt." It was a good business for a young teenager, making $300 to $400 US a month. Like any entrepreneurial petty capitalist, he tried to minimize his labor and increase profit margins by culturing the nightcrawler in similar ways to the red wiggler. But their numbers wouldn't grow. "The nightcrawler will only spit out one egg after it mates; it only has one baby. Red worms [wigglers] . . . they could spit out twenty eggs at the high end. They're in two different ballparks." Jim gave up both picking and growing the nightcrawlers and by the age of sixteen was driving to the local airport to pick up hundreds of thousands of nightcrawlers, air freighted directly from Toronto. To understand why Hugh Carter and Jim Shaw grew the inferior red wigglers instead of the superior nightcrawlers, it is necessary to look a little closer at the family that we colloquially refer to as *earthworms*. Earthworms—that is, worms that live in various layers of soil, as opposed to the parasitic worms in our gastrointestinal tracts, the nematodes in our gardens, or the bristly polychaetes in the sandy ocean coasts—have diverse physiological and ecological characteristics that affect their potential for capitalist commodification.

Broadly speaking, earthworms are divided into three eco-physiological categories based on their behavior, niche in the soil, and feeding and burrowing behavior.[4] *Epigeic* earthworms—from the Greek "upon the earth"—live within and feed on leaf litter and other organic matter at surface soil horizons. Their burrows are ephemeral and would only be used when adverse environmental conditions initiate diapause.[5] They are generally smaller and more reddish, and quickly reproduce. They have a higher metabolism and their rate of ingestion is fast, and they are thus capable of quickly colonizing and digesting organic matter into nutrient-rich casting. Remove feed sources, however, and their populations plummet.[6] *E. fetida* and *Dendrobaena veneta* are both epigeic worms whose economic value as commodities comes through the bait market as well as residential and industrial vermicomposting schemes.

Endogeic worms—"within the earth"—reside in the subsoil, and are geophorous, eating soil and other decomposing matter surrounding plant roots. Their niche and subterranean horizontal burrowing patterns are more challenging to study and thus have elicited fewer scientific analyses compared to the other two earthworm eco-types.[7] They have never been used as commercial bait, and thus I will also dutifully ignore them as well.

Anecic worms—meaning "from the earth"—are species that dwell in permanent burrows as deep as two meters. They come to the surface to feed on

TABLE 1 Characteristics of three species of bait worms

	Eisenia fetida (red wiggler)	Dendrobaena veneta (big red)	Lumbricus terrestris (nightcrawler)
Length, adult (mm)	50–100	50–150	100–250
Width, adult (mm)	4–8	5–8	6–10
Weight, adult (g)	0.55	0.92	4.5
Time to sexual maturity (days)	21–30	65	110–120
Number of cocoons produced (per day)	0.35–1.3	0.28	0.1
Incubation time (days)	18–26	42	70–103
Worms hatched (per cocoon)	23	1.1	1
Optimal developmental temperature (°C)	25	15–25	15

SOURCE: Butt, Frederickson, and Morris (1992); Butt, Frederickson, and Morris (1994); Butt and Lowe (2011); Dominguez (2018); Lowe and Butt (2005).

available organic matter or draw it into their burrows for later consumption. Anecic earthworms also "cast" their nutrient-rich excrement at the surface, with some species, like *L. terrestris*, compiling their waste into tiny towers, or *middens*, that are visible adjacent to their burrow entrances. Mating also occurs at the surface, usually at night, with worms seeking out a partner in a close radius of their burrows. Anecic species tend to be physically larger, with darker hues and much slower rates of reproduction. Their role in agricultural systems is particularly well-researched, as the large porous burrows they produce improve water drainage and soil aeration. One indicator that farmers use for soil health is the number of visible middens in a square meter of agricultural land. *L. terrestris* is the most prominent example of an anecic worm. Open a textbook on earthworms and you will likely see one drawn on page one.[8]

There are primarily three species of earthworms currently sold as live bait. Table 1 details the physiological characteristics of *E. fetida* and *D. veneta*, two species used for both bait and vermicompost, and *L. terrestris*, which is commodified solely for the bait market.

E. fetida, or the *red wiggler*, is the smallest bait worm, but what it lacks in size and weight, it makes up in fecundity. Under controlled conditions, it can produce a cocoon each day, out of which over twenty worms may emerge. The short incubation time and days needed to reach sexual maturity establish conditions for exponential growth. The primary drawback in terms of production, however, is the rather high optimal developmental temperature. Producers can stray little from 25°C (77°F) without significantly affecting reproduction rates.[9] The red wiggler's size and mass also make it the smallest bait worm for recreational fishermen. Though Hugh Carter used to take mail-in orders from bait shops around the country, most contemporary red wiggler farms in North America now produce vermicompost.

D. veneta is a newer worm species that is targeting the bait market. Often called the European nightcrawler, the Big Red Worm, or Dendrobaena, *D. veneta* is sort of a Goldilocks worm commodity, combining the advantageous characteristics of both the red wiggler and the nightcrawler. It is a larger worm, capable of growing up to 150 mm, potentially making large *D. veneta* bigger than small nightcrawlers. But as an epigeic worm, it lives and reproduces on the litter layer and thus can be cultivated in densities closer to those of the red wiggler. Its cocoon production and number of worms hatched per cocoon are much more like those of the nightcrawler, but it does reproduce faster because it only takes about half as long to incubate and reach sexual maturity. Growers in the Netherlands have managed to efficiently produce and export this worm as both a bait worm and a composting worm and are cutting into the nightcrawler's North American market share—a new dynamic that I discuss later.

The capital investments and labor regimes for these epigeic species are diverse. They can be cultivated outdoors or indoors depending on the regional climate, which means limited or substantial capital investments. Hugh Carter's worm beds were constructed outdoors with cinder blocks. Other producers invest in massive tumblers and automated "continuous flow-through" systems where the worm beds and sifting mechanisms are integrated. The cost of such a system ranges according to size. In turn, the organization and composition of labor vary depending on the capital investment. Small and medium-sized farms are commonly run with familial labor, hiring temporary wage labor during specific times if necessary. Worm farms do not require any special skills and thus are encouraged to search out cheap sources of labor—migrants, correctional inmates, and temp agencies.[10] The largest *D. veneta* producer in North America is likely Silver Bait, which operates a highly capitalized production facility with over 10 acres of indoor worm houses, primarily supplying big-box retailers such as Walmart. The vast production facility houses hundreds of millions of *D. veneta* but employs only fifteen full-time employees (including administration and marketing and sales staff), hiring minimum-wage temporary workers in peak seasons.[11]

What do these laborers do on an epigeic worm farm? A United States Court of Appeals case involving Silver Bait provides an excellent description of the capital intensity of the business and how labor is structured more around the machines than the worms themselves:

> A machine lifts the worms and their feed mixture out of the beds and sifts out some of the dirt. Then another machine sorts the worms by size, and workers bring the worms to [the] packing room. Silver Bait makes its own customized bait cups using an injection-moulding machine. . . . Workers in the packing room place roughly thirty worms in each cup. The cups are then placed on a conveyor belt, where they are filled with dirt and fitted with lids through an automated production process. Workers then take the filled cups, add labels, and load pallets for pickup by a delivery service.[12]

To summarize, epigeic worm farmers can better control the ecological conditions of production, which in turn shapes the labor process and the demographics of laborers.

Based on table 1, we can also infer that culturing *L. terrestris* will be a challenging business venture. *L. terrestris* will produce a mere three or four cocoons a month, each needing to incubate for 70 to 100 days—at least five times longer than *E. fetida*. Once hatched, they require another 110–120 days to reach sexual maturity, as opposed to *E. fetida*'s 21–30 days. And after all this long wait, only one worm will hatch from the cocoon. Whereas *E. fetida*'s reproduction is exponential, *L. terrestris* numbers increase on a more linear trajectory. More important, and not captured in the table, is *L. terrestris*'s aversion to high densities. *L. terrestris* does not venture into the space of its neighbors, exhibiting a "reluctance to enter the area around occupied burrows."[13] The best-case scenario, tested in a lab, is to have two *L. terrestris* per liter of soil.[14] To scale up indoor production—say, to manage a stock of 10 million *L. terrestris*—would thus require 5 million liters of soil or about 41,000 garbage cans. This, of course, doesn't take into consideration the turnover time either; fulfilling several orders of a million worms would quickly set back reproduction numbers. Additionally, these figures for the nightcrawlers have never been commercially tested, and I suspect that the density would have to be far lower to achieve optimal physiological reproduction and growth rates. Indeed, research has found mean mass is negatively correlated to increased density of *L. terrestris*;[15] trying to squeeze more nightcrawlers into a container of soil will make for a lower-quality bait commodity.

Cultivating *L. terrestris* therefore makes little economic sense. Instead of trying to control the conditions of nightcrawler production, it has always been more profitable to find landscapes with climates, soil types, moisture levels, and agricultural practices that produce high densities. In short, don't try and grow them; pick them from their "natural" habitats.

COLONIZING SPECIES

Canada's earthworms are a result of colonization. In 1670, King Charles created the Hudson Bay Company and granted it a trading monopoly over the Hudson Bay drainage basin—over one-third of Canada's surface area—as the fur trade was a lucrative business for the English Empire. The shallow waters along the ports of James Bay facilitated the loading and unloading of cargo, such as furs, but also required that the ship's ballast—made up of soil and rock from European shorelines—be dumped offshore.[16] Dumping ship ballast was so common across the Maritimes, Hudson Bay, New England, and the St. Lawrence River that some colonial ship captains became irritated with the amount of pollution they would have to navigate upon docking.[17] Significant amounts of soil, rocks, insects, worms, and minerals were compiling along these ports. The strong northeastern tides mixed these European biophysical elements into the indigenous landscape and established an assortment of new ecological relations and niches, essentially facilitating the "Europeanization" of existing flora and fauna.[18]

It was common to see European worm species at these colonial ports. The interior of the country, however, was a different story. In 1890, Ontario's attorney general wanted to evaluate the economic potential of Northern Ontario's mineral deposits. In the *Report on the Basin of Moose River*, author E. B. Borron makes a peculiar observation: "Excepting at the Hudson Bay Company's Posts (where they have been imported) I do not think I have seen a single specimen of the earthworm anywhere between Lake Superior and James' Bay."[19] Borron was correct; the only records of earthworm observations were associated with colonial trading ports, from James Bay to New England to the Maritimes. Scholar John Reynolds summarizes the consensus of earthworm researchers: "The answer to the earlier question seems to be that the *Palaearctic Lumbricidae* have been introduced to North America by

European settlers since 1500 A.D."[20] But if there were essentially no earthworms in Canada between the last ice age and colonization, what were the soils like?

Indigenous peoples living in what would become the worm capital of the world certainly had no issue sustaining themselves without the earthworm engineers. The Anishinaabe, Haudenosaunee, Attawonderonk (Neutral), Lūnaapéewak, and Huron-Wendat hunted in forests that had evolved without earthworms and fished—whether for fun or food—without the Canadian nightcrawler, as neither Canada nor the nightcrawler existed on Turtle Island. Contrary to the idea that Indigenous populations practiced "swidden" or "shifting agriculture" to combat exhausted soils,[21] recent research suggests Indigenous agriculture was much more "intensive" and "continuous," making it both more sustainable and productive and more ecologically sound than that of the later European settlers.[22] Corn provided most of the calories in the region and was planted alongside beans and squash (the "Three Sisters"), minimizing soil disturbance while retaining organic matter and nitrogen in the soil over extended time periods.[23] In the seventeenth and eighteenth centuries, for example, the Haudenosaunee were producing more grain per unit of land area while preserving more organic matter and nitrogen than the European colonists growing wheat, which required the ecologically destructive plow. This was due to the material characteristics of maize—its seed size, low seed rate, robust seedlings, and higher yields[24]—and didn't require any worms.

The land where worm picking occurs is governed by fifteen treaties signed between 1764 and 1923, which were intended to regulate land and resources between Indigenous and non-Indigenous peoples—often involving payment, hunting and fishing rights, and the establishment of reserves. In practice, the treaties (and the process of treaty-making) were marked by a "clash of paradigms" between the colonial Crown and Indigenous peoples. Historical studies note misunderstanding, threats, coercion, fraud, and forms of blackmail in the process of treaty-making. Nor did the Crown respect Indigenous rights present in the treaties. The "shoddy and fraudulent" agreements were also transformed and altered throughout the nineteenth century, always in the Crown's favor, while purposefully stoking division between Indigenous groups.[25] Contemporary land claims by First Nations in Ontario aim to address injustices rooted in these treaties, citing issues such as fraudulent signings, differing interpretations of boundaries, unauthorized infrastructure development, disputes over infrastructure ownership, harvesting rights, development on unceded and unsurrendered land, and the removal of reserve land.[26]

While most treaties were in place by the early nineteenth century, there were few rural settlers in the region. The town of York (soon to be renamed Toronto) had a meager population in the thousands and was also bustling with anti-loyalist sentiment. In response, the British Crown began recruiting people to populate the treaty land with pro-British settlers to dilute anti-loyalist numbers. In 1825, the British House of Commons established the Canada Land Company, which subsequently purchased 2.5 million acres (over half of which was located in Ontario) for $295,000. The company divvied up the land into 200-acre parcels and purposely developed "planned towns" to encourage and facilitate the sale of agricultural land to white settlers. The company set up agencies at the colonial port cities of Quebec City and Montreal as well as in England, Scotland, and Wales, with flyers, posters, and pamphlets promising fertile land, access to credit, and immediate land ownership.[27]

The settlement efforts were a huge success for the colonists, who attracted not only the farmers who were required to "till the soil" but also the crafts-men, artisans, and shopkeepers necessary to create a burgeoning rural population of over 1,000,000. Southwestern Ontario soon became dominated by small farmers and idealized notions of a family farm where settlers with a strong work ethic and a little gumption could prosper by mixing their labor with the land. The rural landscape was radically altered as farmers began practicing European forms of agriculture, intensively tilling the soil and planting "proper" European crops.

The family farm's material construction and discursive idealization were central to Indigenous dispossession, naturalizing the settler farmers as the stewards of the agricultural land. The same could be said about their traveling companion *L. terrestris*, which only ended up in Ontario soils through processes of colonialism and invasion. The role of earthworms at the time was largely unknown, and few farmers would have purposely introduced earthworms into the earthworm-free soil. There are even records in which farmers, assuming they were pests, blamed them for crop blight. Most likely, *L. terrestris* was accidentally introduced through the transfer of plants and soils. Early colonial records from settlers in Ontario never described the presence of earthworms, and the first mention of *L. terrestris* does not appear until 1902.[28] Even after five hundred years, the earthworm fauna above the approximate glacial maxima remained composed "almost solely" of European species.[29]

The Canadian nightcrawler is thus an impostor. It is not native to Canada and is considered an invasive species, having hitched a ride to colonize the soils while its European counterparts did the same, radically transforming environments that had evolved for twelve thousand years without the presence of earthworms. And yet these invasive species have been remarkably useful for farmers in temperate zones, where *L. terrestris* has become, as one government official told me, the "poster child of soil health."

WHAT *L. TERRESTRIS* DO FOR FARMERS

As visible macrofauna, *L. terrestris* are useful bio-indicators for farmers, who can directly see the burrows, middens, and worms themselves. Some farmers can even hear the worms. Darla, a dairy farmer in Ontario, described walking at night in her hay field and hearing sounds: "I would hear, *zip*, like little *zip*, *zip*. And I thought, what is that anyway? And I looked down, and there were what looked like thousands and thousands of worms. They must have been mating. Whatever they were doing, I was coming along, and they were zipping into their holes." Another farmer, Gabe, knows he has good soil because when he walks out on his field at night, the soil sounds like the snap, crackle, and pop of Rice Krispies. Few of the farmers I spoke with knew much about the specific earthworm species in their land, nor did any of them specify the biochemical reactions going on in the digestive tract. But at the same time they were certainly aware (and have been for generations) of the numerous services they provide to their soil, and ultimately their crop yields. Christopher, a poultry and dairy farmer, had vivid memories of his father's passion for protecting worm populations: "He used to carry a shotgun with him when he was plowing years ago 'cause if he saw seagulls eat worms, he'd be so mad, he'd get out his shotgun and start shooting at these seagulls." The general ethos among the farmers I spoke with was the more worms, the better.

Why is this? What phenomenal benefits do these anecic worms provide that would incite a temperate zone farmer to grab a gun and shoot seagulls out of the air?

First, they burrow. *L. terrestris* is a "mega drill earthworm" that makes macropores in the soil up to two meters deep. Deep in their burrows, the worms are protected from the elements and predators, as they pull food from the surface and plug up the burrow openings with available debris. For the farmer, these large channels are hugely beneficial. They aerate the soil, allowing oxygen to reach deeper levels and preventing anaerobic conditions that deny oxygen to soil flora and fauna, which is beneficial to agricultural production. The burrows also allow water to drain deep into the soil horizons, preventing pools of water at the surface, especially in clayey soils. Studies show how water penetrates the soil between two and ten times faster with the presence of earthworms.[30] This also reduces surface runoff associated with soil erosion and flooding.[31] In addition to filtering water and aerating soils, the pore spaces also decompress the soil to facilitate easy root expansion.[32]

L. terrestris also play a significant role in soil formation, aggregation, and nutrient availability. Their long cylindrical bodies are well-designed digestion machines—mixing organic and inorganic particles and rendering nitrogen, phosphorus, and potassium into available forms for plant uptake.[33] Their castings also substantially increase soil microbial activity,[34] increasing available nutrients to other soil flora and fauna, including the roots of farmed crops.[35] *L. terrestris* burrowing behaviors bring soil to the surface, mixing organic and inorganic materials and mineral particles while leaving a calciferous residue excreted from their skin that solidifies burrow walls, creating rich spaces for material and fungal growth that aids in mineralization.[36] Combine all these behaviors and biophysical processes of this invasive species, and the farmer has healthier or more fertile soils, and thus increased yields.[37]

WHAT FARMERS DO FOR *L. TERRESTRIS*

Changes in agricultural practices, organic and inorganic inputs, and land management will impact worm populations. We might reasonably assume pesticides, herbicides, and fungicides to have deleterious effects on earthworms. And at high concentrations, this is true, with benomyl being perhaps the most toxic.[38] However, earthworms are notoriously good at surviving chemical contaminants lethal to other organisms. In *Silent Spring*, for example, Rachel Carson notes how earthworms did not always succumb to high concentrations of DDT, instead accumulating it in deposits in their "digestive tracts . . . their blood vessels, nerves, and body walls."[39] As a result, these worms with their DDT-infused cells became "biological magnifiers." After eating eleven contaminated worms in a few minutes, Carson noted, robins would drop dead.

What about contemporary chemical inputs? Glyphosate is a wide-spectrum herbicide most prominently known as Roundup. Farmers regularly spray glyphosate as a part of their planting rotation to suppress weeds. The genetically modified seeds they plant are called "Roundup Ready," which means only those seeds will germinate, while other plants are destroyed by the glyphosate. In recent years, concerns over the safety of glyphosate for humans have been reflected most prominently in 2015, when the World Health Organization classified the chemical as a "probable carcinogen." Courts in California have ruled in favor of plaintiffs who claimed their non-Hodgkin's lymphoma was caused by their close contact with Roundup. Monsanto, the manufacturer, has been ordered to pay hundreds of millions of dollars in damages, with appeals ongoing.[40]

For all the important debates on human health, the impact on earthworms is less clear. Under controlled conditions, there have been several studies that conclude even the recommended dosages of glyphosate negatively impact *L. terrestris* surface activity and reduce the mass of both the earthworms themselves and their castings.[41] Other greenhouse experiments, however, find no short-term impact on *L. terrestris* survival, mass, or cocoon production.[42] Field trials suggest the impact on earthworms is more dependent on the rate and length of application, dosage, and methods.[43]

Inorganic fertilizers appear to have an even more negligible effect on earthworms. In some instances, combined with other agro-practices, inorganic nitrogen fertilizers can increase worm populations compared to fields with no synthetic fertilizer treatment, as higher-yielding crops will produce more organic matter for the worms to consume.[44] Agricultural fields that received organic fertilizer (specifically manure), however, had higher worm populations than those applied with inorganic fertilizer.[45]

When tillage and crop rotations are held constant, organic agricultural fields tend to have higher worm counts with greater mass than those fields using inorganic inputs.[46] Organic agricultural practices such as intercropping, mulching, and adding organic matter are particularly beneficial for *L. terrestris* compared to other common earthworm species (though the mechanism behind such rapid increases still needs to be explored).[47] In her excellent and accessible book on earthworms, *The Earth Moved*, author Amy Stewart also describes the auxiliary effects of organic agricultural practices on earthworm populations, noting, "In some ways, it seems like every organic farmer is a worm farmer."[48]

There is a rather gaping caveat to the above statement. Organic management practices appear to have higher earthworm populations when compared to conventional fields, *all things being equal*. In practice, however, all things are not equal. And the thing that affects anecic earthworm populations in farmer fields more than anything else is the plow. Without the ability to control weeds using glyphosate, many organic farmers (though certainly not all) require some form of tillage to suppress weeds. The classic moldboard plow—especially the more modern iteration designed by John Deere in 1837—is a hefty chunk of metal with curved discs that cut deep into the soil. The benefits of tilling soil are immediately apparent. It uproots and kills weeds, disrupts pest habitats, fluffs the soil and creates suitable seedbeds for planting, allows for uniform germination, exposes organic matter to the air and speeds decomposition of leftover crop residue, quickly warms soils after winter, and aids with water drainage. The drawbacks of plowing, however, are more long-term and wider in scope. Exposing massive amounts of topsoil increases rain and wind erosion, the latter of which was evidenced by the 1930 Dust Bowl. And while the topsoil layer is broken up into finer textures useful for seedbeds, aeration, and water filtration, subsoil compaction continues unabated, limiting root development and agricultural yields.[49] The impact on soil organisms is perhaps most devastating, as the plow rips through mycorrhizal fungi and beneficial bacteria that transport nutrients to plant roots in exchange for extrudes. David Montgomery writes, "it is perfectly clear that of all our world-changing inventions, the plow was, and remains, one of the most destructive."[50]

The plow creates all sorts of problems for *L. terrestris*. Conventionally tilled soil for the production of cash crops tends to have the lowest number of worms.[51] Tilling can directly kill the earthworms and expose them to predators. It destroys their burrows and incorporates their future food sources (crop residues) into the soil instead of leaving them on top. Tilled soil also experiences vicissitudes in temperature, particularly under frost in the late fall, disrupting cocoon incubation.[52] More surface temperature variation will also produce uneven moisture levels and affect total biomass.[53] The total impact on tilling will depend also on the intensity (i.e., depth) and frequency. Tillage can reduce earthworm populations by 60–70 percent, although most populations can rebound after a year if intensive soil disruption is not repeated. A general rule of thumb, however, is the following: "The less intensively the soil is disturbed, the less harmful tillage is for the earthworms."[54]

This is a big reason why the untilled pastures of southwestern Ontario have such high earthworm populations. It has long been noted that earthworms are especially abundant in permanent pastures and grasslands compared to other temperate zone ecosystems. As pastures for grazing cattle are rarely (if ever) tilled, *L. terrestris* can reach their maximal density in a given soil type and climate. Pastures also have the added benefit of cattle droppings, which provide a nutrient-rich food source for the worms, increasing both their size and density.[55]

FROM FAIRWAYS TO PASTURES

This is useful information for a nightcrawler capitalist. By the 1970s and into the 1980s, there were significant problems procuring nightcrawlers from golf fairways. First, there were only so many golf courses to go around. It was difficult to expand production when all available golf courses already had leases with coolers or had refused to engage with the industry. Another problem was the increasing rates of herbicide applications. As one greenskeeper who managed several public golf courses in the 1980s told me, 2,4-D herbicides—already a strong chemical—were eventually replaced with Killex, a combination product containing the potent chemical dicamba. Though it is not fatal to earthworms, pickers noticed fewer worms, and those were smaller in size and darker in hue. Coolers paid less money, if any, for these "sick" worms, afraid they would not survive in their flats over the winter months. Crew chiefs, in turn, were not willing to pay as much to access the golf courses at night. By the 1980s, rental rates declined to $1,000 per course, making the arrangement less attractive for golf course managers and greenskeepers who could no longer overlook the problems with garbage, sawdust, and complaints from neighboring houses about the "creepy" headlights at night. Pastureland, by contrast, was bountiful. The perennial grasses and grazing cattle provided an undisturbed habitat and a hearty manure diet for *L. terrestris*. But now they would have to talk to the dairy farmers.

Accessing a piece of property for production is dependent on an array of institutions, sociopolitical relations, and discursive elements. Access, in short, is achieved through a "bundle" of power relations and not solely based on property rights or legal ownership.[56] Talking to greenskeepers and golf course owners was one thing—often, these were private or public corporations that didn't mind losing worms or having people walking on the long fairways throughout the night in a simple exchange for cash. Convincing a farmer to rent land was different, especially for the largely immigrant nightcrawler capitalists navigating racial discrimination in rural white farming communities. It would certainly be easier if access were to be determined by a formal cash-payment stipulation that specified picking hours or dates or the number of worms taken. But it cannot be. The nightcrawler capitalists' ability to access (and maintain access to) land is dependent on the structure of dairy farming and agricultural practices, as well as the discursive understanding of how worms impact the soil. Their ability to access the conditions of nightcrawler production is also determined by the worm itself—its slow maturation, long cocoon gestation, large spatial requirements, and nighttime behavior. There is an "ecology of access" that facilitates and/or constrains the ability of certain people to benefit from specific resources.[57] Finding ways to access a farmer's pasture thus must account for a constellation of social relations, prejudices, market prices, and discursive beliefs, all of which influence how a nightcrawler capitalist can access land to get the worms out of the ground.

The dairy farmer's first contact with the industry usually began with a knock at their farmhouse door. These rural farmers in a homogeneously white region of Ontario were often skeptical. Some of the farmers I spoke with struggled to find the politically correct words to describe the people who had shown up at their doors speaking broken English and wanting to take worms out of their ground. In the 1970s, they "might have been Polish or Portuguese or something like that," said one dairy farmer, or maybe "Greek or Italian," said another. By the late '80s, the frequent knocks started coming from a new group of people, said another farmer, "who don't look like you or me," assuming that they were from Vietnam, Cambodia, or Laos.

The initial rate to rent a farmer's pasture for worm picking was often minimal and always informal, often negotiated on a nightly, ad hoc basis. Some farmers let pickers on their fields for a night in exchange for a case of beer, a bottle of whisky, and sometimes a bit of cash. Scotty, a farmer in his thirties, remembered his father would have worm pickers in his fields for the odd night. "I remember [the crew chief] the most," he said. "If he was alone, he'd always bring Chinese food from his restaurant. We'd be in bed and get out of bed and have a big Chinese feast. We used to love that. We loved worm pickers as kids."

When cash was involved, it was typically exchanged on a per-night basis. The crew chief would show up with a picking crew and pay the farmer between $50 and $200 for a night or two of picking. As a teenager, another farmer, whom I will call Gabe, remembered seeing $200 pass between the crew chief and his father, "and you'd think, 'this is just awesome!'" This was not an insignificant sum of money; land rental rates at the time could be as low as $50 per acre for the year. After ten nights renting to a picker, another farmer told me he made as much cash as if he had rented the land to a neighbor for a year. And by choosing the pickers, he could still use the land to pasture his cows.

But soon these informal arrangements started to get confusing, as demand for land increased. "I rented [the pasture] for the year, and that's when the trouble started," said another farmer, Kurt. After he was paid, he never had to supervise the picking in any way: "You see them out there, they're there, I don't know who they are. I didn't care. I got my money. Go for it." At 3:00 a.m. one night, however, he got a knock at the door from a police constable. Apparently, another picking crew was stealing worms from the crew chief who had initially paid him. It was up to Kurt to sort out which person had legal access to his land. It was even more confusing for Scotty's dad when he found out the crew chief who had rented his land turned around and re-rented the land to another crew chief. With a chuckle, Scotty explained how that initial crew chief would "buy from us, then sell it to all these other guys and make a fortune on it. Dad didn't know who was who then."

Competition between crews was fierce at the time, and trespassing was common. Farmers would often call the police, who would either show up late or simply scare the pickers off with their lights. Even if the pickers were charged with trespassing, the fines were minimal compared to the potential haul of worms, making the risk of trespassing worth the reward. In 1989, a farmer named Paul Maguire managed to get the police to his land while the pickers were still there. The police charged a picker $52 for trespassing. The article notes Maguire found twenty-four mesh bags of worms on the ground, which at the time would have been worth about $240 to the picker. Irate at the police, Maguire didn't understand why the theft of his worms could not be added to the charge of trespassing.[58] Indeed, Alan Tomlin, the "worm guy" with Agri-Food Canada in the 1980s, was often called to the witness stand to clarify the monetary value of worms for these types of cases. "The case that had to be made for the judge was that these had real value," Tomlin told me. In the Maguire case, the court agreed that the value of a single worm would be $0.035.

Trespassing was dangerous for the pickers as well, as they didn't have control over the fields they were going to, nor were they privy to the agreements between the crew chiefs and the farmers. In 1986, another group of Vietnamese pickers were left in a field, completely unaware that they were technically trespassing until the police arrived. In another case, a farmer grabbed his shotgun and fired it in the air to scare off the trespassing pickers.

The farmer's son managed to catch one of the pickers and proceeded to tie him up and beat him, causing a probable concussion and neck injuries. Trespassing charges were laid against some of the pickers, along with careless use of a firearm for the farmer and assault charges for the son.[59] Another older farmer I spoke with remembers his neighbor calling the police and telling them to "get there quick because I'm going out with my shotgun." He wouldn't have shot them, the farmer reassured me. "Just scare them off."

These stories permeated farming communities, leaving many farmers suspicious and distrustful of the industry. "My first recollection was I was fearful of them," Lydia recalled of her feelings as a child. Her father-in-law explained to me how their neighbor found people in his field at night: "He wasn't aware of what it was. They refused to leave. He walked up to one person and touched him to get going and . . . they all pounced on him and held him down, shoved worms in his mouth, and [he] ended up in the hospital. And he didn't live long after that." The father-in-law was adamant: "They contributed to his death. But there were no charges ever laid, because how can you prove what happened out there in the field?" This story spread throughout the community, effectively shutting out the nightcrawler industry from farms in that region for decades. Worm picking only restarted in the small community when a Dutch immigrant—unaware of this decades-old story—decided to rent his fields to pickers.

FROM PASTURES TO ALFALFA

Pasture fields both contained more worms than golf courses and greatly expanded the available picking land. However, the material reality of the new system was also not ideal for maximizing productivity. Pastures were topographically uneven, with thick blades of grass jutting out of the ground next to freshly grazed stubble. Spotting the worms was much more difficult than finding them on open, flat, and finely manicured fairways. Worms enjoyed breaking down the cow patties that dropped sporadically around the picking fields, but this also meant that worm densities were inconsistent throughout the fields. "We'd find a bunch of worms in one area, then have to walk 15 meters until we found another bunch," a Greek former worm picker told me.

In the 1990s, however, the changing dynamics of farming (and dairy farming in particular) altered the ecological conditions of nightcrawler production, which unintentionally increased worm size and density.

Most important—from the perspective of the worm and other soil organisms—was the rapid, widespread adoption of no-till and reduced-tillage practices. In this kind of agriculture soils are never (or less frequently) mechanically disturbed. Crops are directly seeded into the stubble of the previous crop, allowing soil structures to remain intact and soil flora and fauna to develop. The stable subterranean ecosystem enables these organisms to feed from the soil organic matter and maintain nutrient cycling, while the crop residues left over from the previous harvest remain on the surface, protecting the soil from rain and wind erosion. The thick mat of organic matter also retains water and prevents runoff and pollution in nearby waterways, while shielding soils from heat and drought conditions. In 1990, only 6 percent of Canadian farmers practiced no-till farming; by 2021, this had risen to over 60 percent.[60]

For anecic earthworms such as *L. terrestris*, no-till agriculture is a godsend. A global meta-study spanning 65 years in over 40 countries concluded that no-till agriculture increases earthworm abundance by 137 percent and biomass by 196 percent compared to conventional plowing.[61] The increase is more pronounced in temperate soils, with *L. terrestris* as the primary beneficiary. Field trials in southern Ontario also confirm this result. Researchers compared earthworm abundance in fields transitioning from conventional tillage to no-till as well as those converting from no-till to conventional tillage between 1997 and 2012. The results show increases in worm populations with fields converted to no-till, with the greatest reduction in earthworms occurring when no-till fields transitioned to conventional tillage. Fields that had been no-till since 1983 continually maintained the highest populations of *L. terrestris*.[62] This also explain why pastures maintain such high counts.

But cows grazing in pastures are largely a bucolic anachronism. In the 1990s and 2000s, dairy farmers transitioned from outdoor pastures to "confined feeding operations" which allowed them to closely monitor their cows, keep them clean and dry, and, critically, control the quality of feed. By converting the pasture fields to high-yielding alfalfa and corn crops, they could also reduce the land area devoted to feed while simultaneously increasing the quality of feed. Combined with rising land values, this leaves the farmer with a simple calculation. As one third-generation dairy farmer told me, "Do you want to pasture your cows . . . or do you want to make $500 to $600 off of those acres" by growing corn, soybeans, and alfalfa?

When cows move indoors, nutrient-rich cattle manure is not directly returned to the fields. Instead, barns are cleaned out and manure (often slurries mixed with straw bedding) is stored in a manure pit until it is returned to the field with a tractor-pulled manure spreader. For *L. terrestris*, such manure is like an all-you-can-eat buffet spread evenly over every inch of their habitat. Under controlled studies, *L. terrestris* significantly increases in size, mass, and activity with the addition of manure.[63] Christopher was in awe at the speed at which earthworms incorporate the manure into the soil: "Our liquid manure has a lot of straw in it so when it dries on the ground you see a lot of straw residue. So I'm driving along and I can see this line in the field. And I said, okay, I know that's where the alfalfa was, but something looks different. So I got out and looked. There was no manure on the ground. None. It was gone, and here I could see worms incorporated it. So, they say worms will do your tillage for you, if you let them. And I believe it."

Coolers and crew chiefs quickly realized how the changing dynamics of dairy farms were increasing the size and density of the nightcrawlers, particularly on fields of alfalfa, which is a perennial crop; contemporary dairy farmers typically leave alfalfa in a field for three to four years, harvesting it several times a year. The lucky worms living among the alfalfa roots need not fear the tractor plow as long as the farmer leaves the crop in the ground, all while receiving nutrient-rich manure meals spread evenly across the land.

But again, the nightcrawler capitalists still had to adhere to the dialectical relationship between the worms and the farmer. The quantity of land a dairy farmer devotes to alfalfa will depend on farm management strategies, total land size, and rotational schedules. Whether the Ontario dairy farmer owns 100 acres or 700 in total, they will tend to separate their total land holdings into "fields," which can range between 20 and 50 acres each (larger farms may have 100-acre fields). Each field will be incorporated into a multi-year rotational schedule which—for dairy farmers in Ontario—will likely include three to four years of alfalfa (or an alfalfa/hay mix), one year of corn, one year of soybeans, followed once again by alfalfa. Even if the farmer tills the land before planting their crops, the perennial alfalfa fields will not be tilled for three or four years, which subsequently increases earthworm densities. By the late 1990s and early 2000s, it became clear to the nightcrawler capitalists that fields in their final year of alfalfa would possess the most valuable worms and increase picker productivity. Coolers and crew chiefs started to scramble to secure contracts with farmers, and the farmers realized the worms in their manure-laden alfalfa fields were worth much more than a bottle of whisky, a case of beer, or even $200 per picking night. The increased productivity of the worms themselves quickly became capitalized in land rent. If nightcrawler capitalists wanted access to the fattest, most productive worms, they now had to pay farmers for it—a lot.

. . .

The dialectical relationship between worms and agriculture—how the worms affect crop and livestock production, and how forms of crop and livestock production affect worms—shifted the ecological conditions of nightcrawler production away from golf courses to pastures and then to alfalfa fields. Nightcrawler capitalists could not control the production systems through their own design, but were left to seize opportunities and feed off of the dynamism in other sectors to increase productivity. No nightcrawler capitalist intended the increase in worm densities on pastures or alfalfa fields. These changes in the ecological conditions of production happened for reasons completely unrelated to the nightcrawler industry. Nor did farmers reorient production regimes for the purpose of increasing earthworm densities (though they will gladly accept the benefits that accrue as a result of increasing populations). Instead, dairy farms emerged as de facto *L. terrestris* factories in the nightcrawler assemblage through no purposeful action on the part of farmers or the worm pickers. Capital is more opportunistic than deterministic or strategic; the path to further accumulation is not pre-given or necessarily achieved through investing fixed capital in the production process. Instead, the systemic impulse to increase productivity permits experimentation with what capitalists can get their hands on, threading together new human and nonhuman elements (like manure, alfalfa, and farmers) into their existing assemblage while jettisoning the old (golf courses, pastures).

Dairy farmers never intended to profit off nightcrawlers, and the capitalists never wanted to pay them. But through historical contingencies, the nightcrawler capitalists came to depend on the dairy farmers, and a new rent relation emerged slowly, turning the banal worm into the region's most valuable crop.

THREE

Cash-Cropping Worms

There's never less than $1,000 in my wallet thanks to the worm picker.

WALTER, Ontario dairy farmer

HARRISON IS A DAIRY FARMER with 110 milking cows along with 130 heifers, and a sprawling farm with three large barns, each equipped with rooftop solar panels that keep his energy costs to less than $500 a month. The three vertical silos are taller than the pine trees that line the gravel laneway leading to the large tractors that sit between his house and the barns. When Statistics Canada calls Harrison to conduct crop surveys—asking which crops are planted on how many acres—he has a lot to account for. He told me about one time he was on the phone with a surveyor, listing each parcel of land and its crop: 100 acres of bush, 200 acres of corn, 130 acres of wheat, and 150 in soy. He rents out another 80 to a neighbor and recently purchased another 40, which he is renting to another neighbor. Harrison remembered a pause on the other end of the line before the surveyor responded, "This doesn't add up." He was puzzled. "That's strange," he said, "How can I be out 150 acres?"

Then he remembered, "Oh shit, the worm pickers."

Harrison is first and foremost a dairy farmer, but he is arguably also the closest person to a nightcrawler farmer I found. It is not because the worms make up the majority of his crop sales. Nor do they take up much land (as he is prone to forget the acreage when he completes crop surveys). But he is deeply engaged with the industry and incorporates worm picking into his farm management more than anyone I met. "We thought it's a way of diversifying. Because we have dairy, it's stable, but we also cash-crop." Cash crops, he said—whether corn, soy, or wheat—are vulnerable to interlocking vagaries of weather, pests, disease, and the market. "The more diversified you can be, the less risk. And they [the worm pickers] paid good money." Like all the other farmers I spoke with, he had never planned to rent his land to worm pickers. He just happened to possess the right soils, the right tillage practices, crop rotations, and manure applications that turned his fields into ad hoc nightcrawler factories. And nightcrawler capitalists noticed. They tracked the nightcrawlers from golf fairways to untilled pastures, and now to contemporary dairy farmers like Harrison.

But there are a few problems. First, nightcrawler capitalists need dairy farmers, but the dairy farmers don't need them. Canada's dairy industry, much to the chagrin of its international trading partners, is a highly protected industry that guarantees farmer incomes. Why would a farmer even want to "mess around" with this underground industry and have dozens of pickers on their land when they have no need for extra money? Second, won't removing earthworms—the poster child for soil health—negatively impact the land? Third, if a farmer does decide to rent their land, what should they charge? What is the value of the worms in the market, and how does that compare to their value remaining in the ground?

Nightcrawler capitalists have limited control over the conditions of production. Capital must enroll, inherit, or assemble a plurality of actors and actants who may not want or need to be engaged in the industry. I describe the dairy farmers as rentiers, not because they share characteristics with the maligned landed property class of classical political economy, or the greedy rent-seeking corporations and hedge funds of today, but because that is the role the bait industry has thrust upon them. Harrison is an accidental rentier. He is a well-to-do dairy farmer operating in a supply-managed industry that guarantees him a high, stable income. He does not need worm rents; he accidentally stumbled on a valuable commodity embedded in his land through an assemblage of human and nonhuman actors that are spatially and temporally aligned. Now he can make more money. A lot more.

The nightcrawler industry provides a contemporary but peculiar case study to explore, examine, and develop theoretical insights on the emergence and constitution of capitalist ground-rent relations. I begin this chapter by discussing what rent is, how it is connected to capital and labor, and, critically, how it relates to control over the ecological conditions of production. Rent is often called a social relation, but as we will see, it is also an ecological relation. I then document why some farmers engage with the industry and others prefer to "keep the worms in the ground." There is little knowledge about the removal of earthworms, nor is there any way for a farmer to put an economic value on the earthworm's ecological services. I then describe the various worm rental arrangements and the huge sums of money at play to suggest that Canadian nightcrawlers are the most valuable crop in the region, if not the country. With such big checks—and bags of cash—on the table, it is somewhat puzzling why more farmers do not engage with the industry. I suggest the structure of Canada's supply management and the economics of alfalfa production limit farmers' ability to devote more land to picking. The biggest worm rentiers commanding the highest rents, therefore, are those farmers who have more land than they need for their dairy production. I conclude by noting another unexpected benefit beyond the "good cash"—and one that runs counter to everything we thought we knew about worms: removing millions of worms might improve soil fertility.

WHAT IS RENT?

All ground rent derives from monopoly power over exclusive forms of property. It is a social relationship between those who own particular assets and those who want to use those assets as conditions of production—that is, a relationship between landowners and capitalists.[1] Classical political economists generally viewed rentiers as parasitic, siphoning value generated in production while adding no value in the production process. Similar critiques extend to contemporary capitalism, often associated with neoliberalism, where accumulation comes not through producing "stuff" but through speculative land deals, patents, branding, reputation, and finance. Mariana Mazzucato summarizes the difference between the value "made" in the production process and the value "taken" through forms of rent; there are entrepreneurial makers and parasitic takers.[2] In this framing, the unproductive bad capitalism or rentier capitalism can be rectified by the promotion of productive good capitalism, as if getting rid of the rentier will allow the innovative industrialist to flourish.

Such framing around good or bad capitalism, however, can ignore the source of value, as discussed in the introduction. Rents only appear "to generate streams of value out of innovations and/or scarcities," when in reality it comes through the exploitation and control of labor.[3] Rent payments do not come out of thin air but are imbricated in the exploitation of labor and taken out of the surplus value created in the production process. Rent is not a thing but a class relation that manifests itself through the circuit of capital.[4]

At the same time, rent affects how production is organized and what technologies are used, and critically shapes how much of the surplus can be retained by the laborers themselves. Marx, who certainly shared similar disdain for the parasitic nature of the rentier, also understood how rent acts as a regulator and facilitator of capital flow toward efficient production processes, "forcing the proper allocation of capital to land."[5] It can equalize competition by compelling the capitalist to pay for certain so-called natural or monopolistic advantages that stem from the specific qualities or location of the land. By taking a slice of the surplus, the landlord taxes away what might otherwise be "free gifts of nature" to one capitalist, thereby equalizing and stabilizing the terrain between competitive capitalists.

This leads to another implication, however banal or retrospectively obvious: ground rent is not just a social relation but a socio-ecological relation that depends on the extent to which nature can be controlled, transformed, or bypassed altogether. In this sense, I suggest rent is tied to both the formal subsumption of nature (e.g., rents for certain types of mines, agriculture, forests, wild fisheries) as well as the real subsumption of nature (e.g., legal enforcement of patents on genetic seeds). Few authors have explicitly developed the relation between rent and subsumption, which is somewhat surprising considering Marx discusses soil dynamics and the impact of industrial agriculture on England's environment, specifically in relation to rent. Carlo Vercellone most explicitly sees a relationship between fixed capital investments and the real subsumption of labor as a means to counter the monopolizable characteristics that gave rise to rent in the first place.[6] In a similar vein, David Goodman and Michael Redclift see the relation between formal and real subsumption of labor as tied to the "organic" non-reproducible aspect of land and capitalist attempts to accumulate capital "off the farm": "In our view, the real subsumption of agriculture is not to be observed at the 'point of production' of the farm. Rather, it is represented by the long-run tendency of capital to eliminate the labor process as a 'rural' or land-based activity."[7] Their concepts of substitution and appropriation are strategies in which capital can avoid the land problem and accumulation can occur off the farm. Robin Murray explicitly connects the magnitude of rent and fixed capital investments, noting how rent arises when capital cannot reproduce the condition of production for its own sake. Once capital can find substitutes for the monopolizable characteristics of nature, say substituting fluorescent light instead of the sun, "the material basis for rent of all kinds is thus dissolved."[8] The problem for agriculture and other land-based industries is that capital "still exercises formal rather than real subordination over the land. This is the material basis for . . . rent."[9]

We can see this clearly with lively commodities and other industries that confront nature directly in the production process. When capital struggles to transform the ecological conditions of production, it must deal with nature in situ and subsequently confront the owner of such conditions for production to occur.[10] Control and design over the ecological conditions of production is essential for understanding the distribution surplus value. Reshaping the *ecological* basis of rent reshapes the *social* relation of rent.

This chapter offers a peculiar case study that documents how control over and redesign of nature are inherently tied to the ecology of rent relations and the subsequent struggle over surplus value. Nightcrawler production must access specific parcels of land because it cannot transform its own ecological conditions. Golf courses led to pastures, which have currently led to dairy farmers. Alfalfa fields and cow manure have drastically increased earthworm densities and the productivity of the pickers themselves. But this does not mean pickers get more of the surplus. In this case, the extra surplus value embedded in each worm must go to the farmer, even if they do not wish to play the rentier role.

For Ben Fine, there can be no general theory of rent applicable to all empirical cases, and therefore discussion of rent must always be situated and linked to specific historical-geographical conditions.[11] The experimental nature of capital accumulation and its relation to nature suggests there may be a variety of forms the rent relations may take.[12] Most Ontario dairy farmers *own and operate* the land; they are generally not rent-seeking landlords looking to rent their assets to the highest bidder. Devoting land to worm pickers inherently affects the organization of their own agriculture production while at the same time removing what is thought to be one of the most critical organisms in maintaining soil health. When dairy farmers consider whether to rent or not, they are making more than an economic calculus based on expenses and returns, but also must include their conceptions of the earthworm as a critical component of their own ecological conditions of production. What will happen to their land and soil if they remove millions upon millions of the most prominent ecosystem-engineering macrofauna? What is the value of the earthworms, and should they put a price tag on them?

"THERE ARE THOSE WHO TREASURE THEIR WORMS, AND THOSE WHO DON'T"

I called an organic farmer, Peterson, for an interview about his view of worm pickers. "Oh, this will be quick," he said over the phone. "I would never, ever,

ever allow any worm pickers on my land." He described his soil as a machine with a million moving parts consisting of bacteria, mycorrhizal fungi, nematodes, root systems, decomposing organic matter, and of course earthworms—each one interacting in a complex dance with the others. "How could you take out some of the most important parts?" When I met with him in person, he pulled out a 1947 book by Thomas Barrett called *Harnessing the Earthworm*, a poetically written ode to the earthworm—or the "mill of God"—that is perhaps full of more reverence for the lowly creature than Darwin's own book. Peterson spoke in metaphors to convey the function of earthworms in his holistic ecosystem. Worms are but "one pixel in the big picture," capable of excreting vast amounts of active and stable nutrients that are indispensable to the functioning of his farm. "For us to sell the worms would be like selling the seeds instead of planting them. It's like selling the fuel out of your car and then you walk." As Peterson told me: "I know in an acre of land, the amount a cow weighs is the weight of worms in the soil. They eat and they digest the soil. They eat plants and debris, and enter it into nutrients. For us, it's just a no-brainer that we don't want to sell our worms." His estimates are conservative. Two hundred thousand of the largest, juiciest worms that the worm picker could take from an acre throughout the year would weigh approximately 1800 pounds. An adult Holstein cow? One thousand and three hundred pounds.

Whether we call earthworms the poster child for soil health, the mill of the gods, or the intestines of the earth, few soil macrofauna elicit such universal praise. Farmers want worms in their soil, and the more, the better. The scientific literature, as described in the last chapter, backs such lofty titles and justifies Peterson's praise for the worm. Darla, a boisterous and straight-talking dairy farmer who dabbles in a variety of alternative production methods, cited the effect on soil ecology as the primary reason to reject worm picking. She was explicit: "You really have to understand [the worms] and understand the sustainability of that life-form and figure out how everything works together. So if, for example, you do take the worms out of the fields, then what? What are the ramifications of that? Because we know and see it. They're very good at incorporating surface materials. And we also know they make holes, so the air and water can go down there. We know they excrete materials. . . . I don't see how you win by removing a beneficial thing."

Gabe considered the loss of worms as an unknown calculation: "You have to think about not only . . . their monetary value, but also the value of keeping the worms in your soil. And I just figured . . . those worms are doing more value for me in the soil than they [the industry] are giving me in cash." His perspective was shared by other farmers, especially organic farmers. One prominent organic dairy producer expressed his comments via his secretary: "His comment was that worms are some of the most valuable creatures of the soil and the worms do amazing work. He will not let worm pickers on his land."

Farmers are hesitant because there is really no information available to them. The research on *increasing* worm populations to the benefit of agriculture is clear. There are few studies, however, that document the impact of massive earthworm removal from temperate farming systems. If a southwestern Ontario farmer wants a study that specifies the potential impact of worm picking on their soil health, they will not find one. "That's part of the challenge," said another farmer who has often chased pickers off his land. "As a landowner, I have no idea what the value is of what they're harvesting . . . the benefit of my soil and what they're taking." And many farmers, like Darla, are not willing to take the risk: "I don't know how sustainable the worm-picking industry is. Is it concentrated in certain areas, is it denuding it of worms, do they have a way of reproduction, and can they come back quickly? I don't know. But I do know that any one of these fly-by-night industries that just rise up, there's no control and no research. So we need to get a handle on these things."

I suspect the lack of research is simply because farmers rarely have a reason to take worms out. Agricultural research is often more interested in getting good yields than examining the impact of removing worms. There are only a few studies that would seem to confirm the intuitive drawbacks of removing earthworms. One small study showed how removing anecic earthworms influenced the feeding habits of other species and increased the population of endogeic species.[13] Other controlled experiments demonstrated the inverse of increasing earthworms: decomposition of organic matter was slower with lower densities,[14] as were rates of water filtration and aeration.[15] This much should be obvious. But what about in a real-life setting? When a farmer asks crop advisors or soil scientists about what would happen to their soil ecology if they removed between 200 and 400 thousand worms from an acre of land, they are largely met with a shrug.

None of the soil scientists, earthworm experts, and government officials I spoke with were too concerned about picking—not because they had scientific evidence to back it up, but because they reasoned the impact would be minimal. Alan Tomlin told me how overpicking one's worm population would be difficult. With the pickers only interested in the large worms, "you probably reduce the size and average age of the population to some fairly stable point and then you just wouldn't get them anymore." Tomlin suggested seagulls following behind a plow would probably be more efficient worm pickers than humankind, and even then, the impact is negligible: "We actually ran some tests. . . . They [gulls] just don't make a dent in the population [of worms]." Crop advisors in the worm-picking regions of Ontario are often asked the same question. While they are self-admittedly no worm experts, they don't see a reason for farmers to abstain. One crop advisor I spoke with said moderation is key. "My argument is always—the health of the soil is paramount for longevity. Yeah, okay one year, maybe, maybe not." OMAFRA suggested the same thing: maybe do it for a year, and it probably wouldn't affect anything. I explained this to Peterson, who came around to the logic. "I can see the concept . . . when you have three years of alfalfa the soil is undisturbed, there's a lot of plant debris that's fallen down. The worm population increases dramatically." But he followed up, "Then you have to ask yourself, what happens if you do it every year?" A fair question—and one that a couple of farmers can answer.

Renting the same fields year after year to pickers is rare, but whether this is due to decreasing worm population or simply because farmers need to get the land planted in crops (see below) is unclear. Harrison was the first person I spoke to who can answer Peterson's question: "Now we're at the point we do it year-round. The field that's still in worm picking now has been in worm picking for probably six or seven years. Hasn't grown a crop other than worm picking. And he [the cooler] still keeps wanting it; he says he wants it forever." To repeat, Harrison has rented the same field to pickers for the past seven years. It is not even part of his rotation. It is a perpetual worm field. He's not particularly interested in the intricacies of how or why this field produces so many worms. "I don't care about the life cycle of a worm," he told me. All he knows is that the cooler always wants the field and would happily accept more land.

Before the transition to reduced tillage, one crop advisor joked that "you could tell the farms that had good soil by the number of seagulls that were following the plow." Plowing fields overturn the soil and expose the hapless worms to the hundreds of gulls that would patiently follow behind the tractors. In an era of no-till farming, however, I would propose a different measure for farmers to gauge earthworm populations: are worm pickers knocking at your door with thousands of dollars asking for the right to pick worms?

Organic farmers like Peterson are actually at a disadvantage. In fact, for all the talk of "keeping the worms in the ground," very few organic farms ever receive these lucrative knocks at their doors. Without the ability (or desire) to apply herbicides like glyphosate, they need to plow fields to overturn plant and weed roots before planting their next crop. There are some new strategies to manage weeds without tilling—such as planting annual cover crops, which are then broken down by a crimper—but most continue to till. Therefore, on the one hand, organic farmers should be very protective of their nightcrawler worms, largely because they will not have as many as a no-till farmer spraying glyphosate. On the other hand, the paucity of anecic worms means they may never attract the attention of the worm pickers in the first place.

I never met a farmer who noticed any decrease in yield after having pickers, or any impact on soil structures. The high rental rates are thus not compensating for some nebulous calculation of the value of the worm to their soil or to offset subsequent yield drags. But there are other drawbacks. What frustrates farmers most about the pickers is the garbage that is left behind. This was a significant issue, especially in the 1980s and 1990s. Farmers recalled a whole assortment of waste items left behind: snack bags, plastic wrappers, gum wrappers, cigarette butts, energy drinks, glass bottles, rubber gloves, headlamp battery packs, and shoes. One farmer saw adult diapers; when he inquired, the crew chief stated this was because pickers don't want to stop for a second when the worms are out in full force. And yet, worms are not out in full force every night, leaving time for other leisurely activities under the cover of darkness. One farmer remembered his father finding a condom in the field after a night of picking. "And that was it [for worm picking]. No more."

Nor are farmers particularly comfortable with having other people walking all over their property. Geoff—who does not have pickers—said that as a farmer, he's always "leery of people coming on your property. . . . We don't want people on our land at all. Anywhere. That's just the way we are. We're very protective of our land." He doesn't care whether it's a worm picker, some teenagers partying, or a passerby grabbing a few ears of corn. "We lose our Mennonite status during our growing season," he said, alluding to the stereotypical pacifist nature of his religious faith.[16] "I think that's the polite way to put it."

Worm picking, however, does compound the discomfort. I suspect few people—whether they are being paid for it or not—enjoy having dozens of people roaming their property with headlamps while they are sleeping in their beds. "It has always been freaky to me to see the lights out in the field," said Lydia, who has no positive memories of the pickers. And they are walking everywhere—"around the barn, by the houses, along the laneway." Gerry decided to stop having pickers when they startled his wife when they were picking worms right next to the bathroom window of their house.

Taken together, the perceived threat to their worm population, the garbage on the field, and the general discomfort with the nighttime encroachers are enough to dissuade many farmers. The rental rate would thus have to account for these drawbacks to make worm picking worthwhile for the farmers.

WORM CONTRACTS

Diedrich was one of the few farmers who had a formal contract with a cooler, which documents the rights and responsibilities of the lessor and lessee. It was a four-page document, written in legalese, complete with a preamble and a line for the witness's signature. The document began, "Now therefore this agreement witnesseth that in consideration of the mutual covenant and agreements herein and subject to the terms and conditions of this agreement, the parties agree as follows. . . ." What follows are the dates on which picking could commence and when it must end. It stipulates who is responsible to "harrow the fields," who will haul manure, how much manure must be applied, as well as who will pay for the Roundup applications. Contracts like these are rare, but it allayed Diedrich's fears over workspace insurance issues and provided some clarity around his rights and responsibilities in the rental relationship. "If it wasn't for a solid contract with certain rules and agreements," he told me, "I don't think I'd be interested." Lena, another farmer renting to pickers, also has a typed contract—but one that she typed on her own computer while sitting with her crew chief. Other types of contracts exist but are typically scratched down on a notebook or a piece of paper. Harrison's contract is a hand-written scribble that states, "I, [cooler name] rent this field at [specific address]." Christopher has no contract at all. "Just on a handshake?" I asked. "Yup," he said. "There are no documents?" I followed up. "Nope." Coleman, another farmer deeply embedded in the industry, also has nothing to show for the arrangement, save the checks he receives from his cooler. He was surprised others did: "Do you have people who actually have contracts? Really? Wow." Below, I use Diedrich's contract as a guide to describe the rights and responsibilities of the crew chiefs or coolers, those of the farmer, and how they influence the rental rate.

FIGURE 2. Diedrich's unassuming worm field. Photo by author.

How Much Land, for How Long

> The Lessor hereby leases the Lessee the lands as indicated above, containing approximately _____ acres (hereinafter called the "Lands") for a term of one season, commencing _____ and to terminate on _____.

The primary determinant of the price of land is simply how long the pickers can access how much land. The size of the field will largely be determined by the farmer's rotation and total land base. There are four broad arrangements that establish the duration: the *last cut arrangement*, the *first cut arrangement*, the *full-year arrangement*, and the *perpetual arrangement*.

In the simplest arrangement, or what I call the *last cut arrangement*, the farmer will rent a field to the pickers after their last cut of alfalfa in September. Why is this simple? Ontario dairy farmers generally plant a field of alfalfa to last for three to four years in total. Each year, it is harvested or "cut" at least three, sometimes four times, and in productive areas (with the help of weather) five times. The *last cut* of alfalfa, however, will usually be completed by mid-September. This means the picker has access to the fields for the

AND WHEREAS, the Lessee operates a bait business having its head office in the city of Mississauga, in the Regional Municipality of Peel;

AND WHEREAS the Lessor has agreed o grant the Lessee the sole and exclusive rights to pick dew worms over the Lessor's Lands and premises, for and during the term set forth;

NOW THEREFORE THIS AGREEMENT WITNESSETH that in consideration of the mutual covenants and agreements herein and subject to the terms and conditions of this agreement, the parties agree as follows:

1. The Lessor hereby leases the Lessee the lands as indicated above, containing approximately ___40___ acres (hereinafter called the "Lands") for a term of one season, commencing __Sept.25, 2017__ and to terminate on __Nov.15,2018__

2. The Lessee covenants and agrees to pay the Lessor:
   For field I: (Pasture) _____ acres, the sum of $_____ per acre, for the total amount of $_____ payable as follows:

       (a) $_____ on execution of agreement
       (b) $_____
       (c) $_____

   For field II: (Hayfield/Alfalfa) __40__ acres, the sum of $__20,000.00__
   Payable as follows:

       (a) $__7,000.00__ on execution of agreement
       (b) $__7,000.00__ July 25, 2018
       (c) $__6,000.00__ Aug. 25, 2018
           $ 20,000.00

   For field III: (Soybean/Corn/Wheat) _____ acres, the sum of $_____
   Per day, per vehicle. Payable as follows:
   $_____ on the execution of this agreement for the first _____ vehicles.

   No Till soybean/corn field must be harrowed to eliminate the debris on same to enable the picking of worms.

   a)The Lessor agrees to apply 7,000 gallons of dairy manure per acre after the field is sprayed and mowed. It is paramount to apply the full 7,000 gallons of dairy manure per acre to achieve the right size of worms to be marketable. A further coat of 7,000 gallons of dairy manure shall be applied in early May,.2018.

FIGURE 3. A rare worm rental agreement. Photo by author.

prime picking months of September and October, which is critical for the coolers who must restock their warehouses and keep the worms over winter to supply retailers at the start of the spring fishing season. Usually, under this arrangement, the pickers also have the right to pick worms the following April and the first week of May—weather permitting—before the farmer plants corn. This last cut arrangement will cost the crew chief or cooler between $200 and $400 per acre. Certainly, this is no cash windfall for the farmer, especially compared to their income from milk. And yet, a 30-acre field could generate an extra $6,000 to $12,000 with no added effort from the farmer. They do not need to change their planting or harvesting, manure

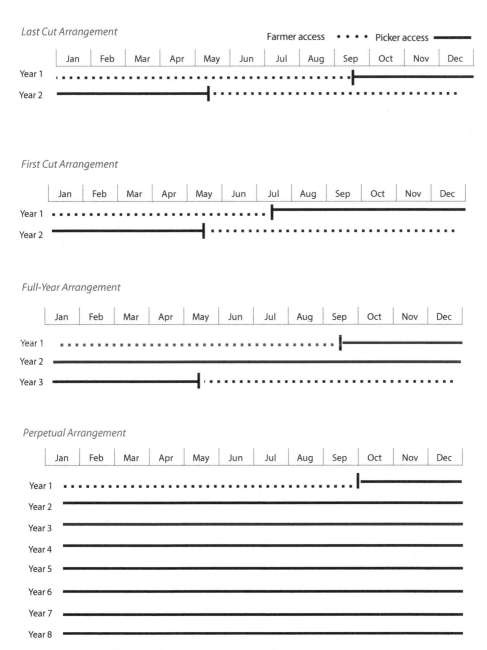

FIGURE 4. Types of worm picking arrangements with farmers.

application, or herbicide treatment. Nor do they need to worry about their feed stores. They will harvest their entire crop and will plant the following spring in the usual way. Although these rental rates are on the low end, the only extra labor expended, as one farmer told me, is to cash the check.

Farmers who are confident in their feed supply may forgo the second or third cut of alfalfa and grant access to their fields after their first or second cut, usually in June or mid-July. This *first cut arrangement* gives the pickers at least two more months throughout the year and thus will pay the farmer between $350 and $600 per acre. Scotty notes the crew chief will pay him $500 if they can have the field after the first cut. After the second cut, $350: "I can't grow near that [dollar amount] in hay . . . so you're like, 'oh boy, we better give it to her.'"

Crew chiefs and coolers, however, prefer to have the land for the full year. In this *full-year arrangement*, farmers formally incorporate worm picking into their crop rotation. The new rotational schedule becomes three years of hay/alfalfa, one year worm picking, one year corn, and one year soy, before returning to alfalfa again. For the pickers, this means they will get access to the field after the last cut of hay in mid-September, retain picking rights for the entire following calendar year, and still pick more the following spring until the corn is planted. The farmer grows nothing else but worms. For the pickers, this increases the number of optimal picking nights throughout the year, particularly the months of May and June, which are otherwise inaccessible because of the farmers' planting schedule. And while the pickers technically get two falls and two early springs in this arrangement, the farmer only forgoes a single year of planting. The financial incentive for the farmer is phenomenal. Farmers receive between $650 and $1,500 per acre, depending on the soil quality and location. Harrison currently gets $1,100 per acre. At 110 acres that's $121,000. Another farmer, Walter, gets $1,200 per acre on 60 acres for a total of $72,000. Both are expanding the amount of acreage devoted to worm picking because . . . well, they can do math.

When I mentioned these dollar amounts to farmers who did not have worm pickers, their eyes would subtly widen. "If you get $1,000 an acre, I can't argue. That's 100% income." Peterson, the organic farmer, was equally surprised: "He gets paid that much?! He couldn't grow a crop for $1,000 an acre." Even a representative from OMAFRA, who called a few farmers before our interview, was surprised: "I was shocked when a dairy farmer told me $1,000 per acre. There aren't too many legal grabs that can net you that."

The final arrangement, the *perpetual arrangement*, is for the few farmers who rent the same field each and every year, like Harrison. For reasons largely unknown to both farmer and picker, these fields continually produce a lot of big worms. Harrison just has to make sure of one thing: manure.

The Importance of Manure

Article 2a) The lessor agrees to apply 7,000 gallons of dairy manure per acre after the field is sprayed and mowed. It is paramount to apply the full 7,000 gallons of dairy manure per acre to achieve the right size of worms to be marketable. . . . For verification purchases, the Lessor shall notify the Lessee after every manure application.

If the crew chiefs had their way, the farmers told me, they would coat the fields in manure any chance they had. In his contract, Diedrich puts 7,000 gallons of manure per acre on his worm fields twice a year. Scotty chuckled, "All she [the crew chief] wants is manure." He sprays four to five thousand gallons per acre three times a year for the picker. "She'll tell me, 'The worms are too small, my buyers are complaining. We need bigger worms; we need more manure'. . . . She would take all she could get. . . . More more more manure."

Most farmers spread between 3,000 and 6,000 gallons per acre several times a year for the pickers. Harrison applies 10,000 gallons per acre, which is likely why his worms keep growing so quickly. But of all the peculiar machinations of the worm-picking industry, "this is the part I'm not too sure about," he said. "I don't know what this does for nutrient management plans [NMPs],[17] all that kind of stuff, because we put way more manure on that field."

Indeed, those who put more manure on for the pickers tend to receive higher rents. Several farmers mentioned a large dairy farm receiving at least $1,500 per acre. "They are just pouring manure on," said Coleman. One cooler told me, "It's really unbelievable what's going on there," pouring on so much manure that it clearly contravenes any NMP. But it turns worms into veritable "snakes," he acknowledged, and snakes are what his American customers want. If his 500-worm flats do not weigh 9 pounds each, his Wisconsin buyer is not interested.

For a while, Harrison never put additional coats of manure on his worm-picking field. He always had a lot of acres and figured he would rather spread it around to fields that could make better use of the embedded nutrients. On average in Ontario, each load of 5,000 gallons of liquid dairy manure contains about 70 pounds of usable nitrogen, representing an approximate value of $100 per acre. Therefore, when Harrison applies his manure to his other fields, he is

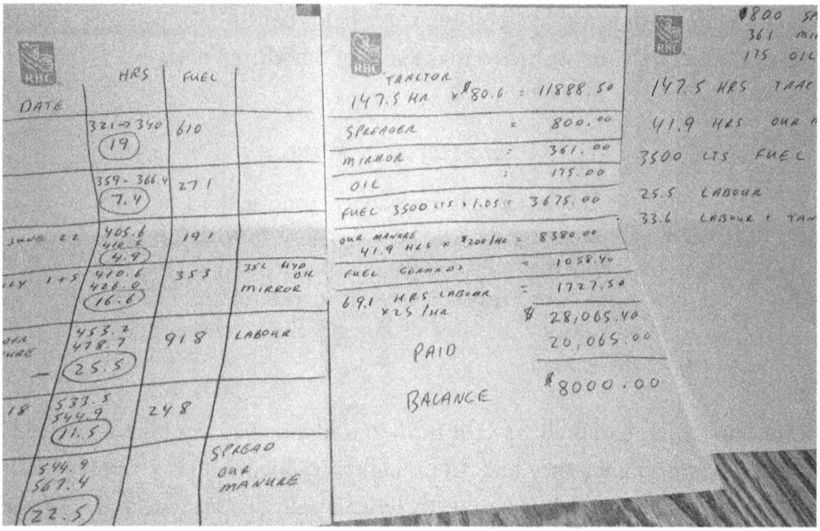

FIGURE 5. Harrison's worm-picking expenses. Photo by author.

pumping both nutrients and their monetary equivalent into the land. The crew chief, however, still wanted more manure on his rented field, and when Harrison didn't put on another coat, the crew chief purchased additional manure from a neighboring farm to apply on Harrison's field. Obviously, Harrison didn't have a problem getting this free gift of nutrients to his land, but he thought if the crew chief was willing to pay for manure, why not sell him his own? "So, we're basically hauling our *own* manure on our *own* land, and they're *paying us* to do it" (emphasis mine). Harrison now requires full payment for all his services over and above the rental rate. He showed me his list of expenses. In 2018, for example, he was paid $90,000 for rent, as well as an additional $28,000 for transportation, manure, fuel, and labor expenses—most of which he would have incurred regardless of his agreement with the pickers.

CASH FOR CROPS OR CASH FOR WORMS?

Scotty gets his bag of cash three times a year. "She comes in with her wads of cash and we start counting," he told me, followed by a laugh. Walter gets all of his cash—$60,000 to $70,000 a year—up front. Then, as a perk, "every single Christmas, he [the cooler] drops me off a bottle of liquor with an envelope of $1,000 cash. Every Christmas. That's my favorite Christmas card." At this point in the interviews, I asked the uncomfortable question: What do you do with the cash? Do you report the income? Scotty claimed half the cash as rental and split the other half with his parents, calling it "spending money." When I asked Lewis the question, his eyes darted to his father, who was sitting across the table from him in their break room. This led to a wry smile and a definite response. "No." They both laughed. I asked Christopher what he does with the "wads of cash" he receives. "It goes into five pockets," he said, as he patted his pants pockets and the breast pocket on his shirt. A similar sentiment is shared by other farmers, who simply see it as a year-end bonus. "Honestly," Lewis said, "it's nice cash."

What exactly constitutes "nice cash"? There are a few ways we could compare worm picking on yearly contracts with traditional farm income. Walter has done the calculations comparing worms to corn and has concluded that, "for me, it's a no-brainer." His fertile land can produce 200 bushels of corn per acre, "which is really good," and can sell at $4 per bushel. At first pass, therefore, worm picking appears only slightly more than the $800 per acre he can get from corn. But then he has to calculate all of his expenses. After the costs of seeds ($300), fuel, spray, labor time, as well as "insurance and all that shit" to minimize risk, Walter considers himself lucky to net $150 per acre from corn. Harrison made similar calculations. He assumes he can gross $1,000 an acre on corn, but his expenses are at least half of that. "You'd spend at least $500 per acre, and then the rest would go to the cost of the land and your management." Even after these operating and capital expenditures, however, farmers still have to devote considerable labor time to manage a perishable product. "You gotta dry the crop and store the crop and all that kind of stuff," Harrison said, which can be a logistical nightmare. Worm picking is much simpler for Coleman as well, when he considers the risk of growing a crop susceptible to disease, spoilage, weather, and market prices. All that work, he said, for a measly profit of $100 per acre.

OMAFRA provides a field crop budget, as well as average yields and crop prices. Table 2 examines only the most profitable area, Oxford County, which has the highest crop yields in the province. These are very generous projections, obviously dependent on region, farm management, total land base, weather, and market prices. No farmer I spoke with, even some of the largest landholders, believed they could net these numbers. Most farmers guessed their net income on corn at between $170 and $250 based on their past year. However, even compared with these exaggerated numbers, we can

TABLE 2 Estimated net income per acre for crops in Oxford County

Crop	Average yield[a]	Price ($)[b]	Gross income ($)	Expenses ($)[c]	Net income ($)
No-till grain corn	184.5 bushels	6.38	1,177.11	776.75	400.36
Silage corn	24 tons	45.82	1,099.68	1,177.15	−77.47
No-till soy	53.2 bushels	14.66	779.91	375.70	404.21
Alfalfa-timothy grass	2.4 tons	180.5	433.2	468.15	−34.95

[a] Information for average yields comes from OMAFRA (2024).
[b] As prices constantly change, I am presenting the five-year average from 2018–2022 from OMAFRA (2024).
[c] Information for estimated expenses comes from OMAFRA (2024).

easily say that worms are the most profitable cash crop in Ontario, and perhaps the country.

But is this a fair comparison? If we do not conceptualize the worms as cash crops and view the income solely through the land rent, how then does worm picking compare to the dominant agricultural land rental rates in the region? If you're a farmer with a few extra fields, would it be more profitable to rent to your neighbors or some worm pickers? Again, from a financial perspective, it's a no-brainer. Even in some of the most fertile land in the country, average land rental rates are significantly less than what the crew chiefs and coolers are offering:[18]

Huron	$363/acre
Perth	$372
Waterloo	$319
Wellington	$198
Lambton	$303
Chatham-Kent	$314
Elgin	$312
Oxford	$465
Brandt	$269
Haldimand	$296

Coleman notes that high-quality land in a neighboring county rents for $400/acre for conventional crops, with worm pickers offering $1,200. In his county, land might rent for $240/acre, with the worm pickers offering between $650 and $850. Diedrich rents his land to pickers for $850 (though the receipts say $500), but would likely receive only $350 renting to neighbors. Even if he were only getting $500 from the pickers, as his receipts say, "you're still getting better, because there's no work. They pay all the expenses too." Harrison could get $350/acre to rent his excess land, but he'd rather take the $1,100/acre from pickers. He is also keenly aware of how to shuffle his land between pickers and neighbors. If, for example, he has just harvested an 80-acre soybean field, he knows the worm pickers will not be interested in renting it. Instead, he will rent the field to a neighbor who may want to grow alfalfa for feed for three years. At the conclusion of this rental agreement with the neighbor, he can take back the land (and the boosted worm counts) and rent it to one of many crew chiefs who will be knocking at his door, ready to pay a premium.

Perceptive readers might have calculated the opportunity for arbitrage in the worm market, something not lost on some of the farmers. On one hand, it makes more economic sense to rent their manure-laden fields to worm pickers and simply purchase the necessary feed for their cattle. If an acre's worth of alfalfa has a market value of a few hundred dollars (a very high estimate) and pickers are willing to pay $1,000, then the farmer only needs to purchase feed instead of growing it and profit close to $800 per acre.

Alternatively, farmers could rent fields from their neighbors and turn around and re-rent them to the worm pickers. For instance, one could rent 50 acres coming off alfalfa from a neighboring farm at $350 per acre, or $17,500 in total. Instead of planting corn on this rented land, they could re-rent the field to worm pickers at $1,000 an acre, immediately profiting $32,500 by simply signing two contracts (or simply shaking two hands). These arbitrage opportunities appeared clear to me. However, farmers are hesitant to potentially alienate their intergenerational neighbors with such a tactic; as one farmer mentioned, "You gotta live beside that farmer for the rest of your life, and you don't want that potential animosity." Harrison remembers one example of a farmer renting land from a neighbor and then immediately re-renting to pickers. The result? "They don't talk anymore."

But then I had another idea that I proposed to Harrison. What if I rented *my own land* to worm pickers for the year and *rented my neighbor's land* to grow my feed? The price differential would be the same, and the additional costs of growing hay on a neighboring field (assuming it's geographically close) would be marginal. That is, take advantage of the low cost of feed and reap rewards from the high rents. Harrison thought about it for a moment:

"No one would have a problem with that," he said, though he didn't know anyone who did this. A few months later, however, I met Coleman.

Coleman does not own enough land for his 160-head herd, neither for feed nor manure (that is, he cannot spread all of his manure on his own land because of the restrictions in his NMP). He can easily sell his manure to other farms in the area but is always looking for nearby land to grow his feed. As a result, he feeds his cows a higher ratio of straw and tends to purchase alfalfa instead of being self-reliant. When he looked at the worm-picking numbers, he decided it made more sense to have pickers on his own land, rent nearby land, and buy additional feed on the market. He has maintained neighborly relationships in this form of arbitrage-once-removed. To me, and apparently to Coleman, the economics seem clear. The nightcrawler is the most profitable crop imaginable, with next to no work or risk. It seems like free money that few farmers are willing to take. Why is that?

SUPPLY MANAGEMENT AND THE VALUE OF ALFALFA

Canada's dairy industry is regulated through a "supply management" system that discourages overproduction by limiting production through a quota system, receiving guaranteed "minimum prices," and restricting imports through tariffs. A wealthy dairy farmer in the United States could increase production by purchasing more cows and producing more milk, often with government subsidies based on acreage or yield. In Canada, there are no subsidies; it is illegal for a farmer to sell more milk than their quota allows. If a Canadian dairy farmer wants to increase revenue they must either buy a more expensive quota (if available) or become more efficient. "Efficiency" was a common refrain I heard from the farmers: "If you can't increase the gross income then you need to be more financially efficient at doing it." Put in more theoretical terms, capping the production of commodity output—like a quota system—means that increasing profit comes only through the logic of relative surplus value, ensuring that the amount of labor time decreases while the working day (or in this case, production output) remains the same. How can farmers hit their production quotas with fewer resources?

Critically, supply management quotas are based not solely on milk volumes but also on the percentage of butterfat the milk contains. That forms the Total Production Quota (TPQ), which equals the total liters produced multiplied by the butterfat percentage in the milk converted to kilograms per day. Bradley explained the details to me more clearly. Imagine, he said, you have a daily quota of 60 kg. "Let's say I ship 3,000 liters of milk every other day at 4 percent butterfat. That would be over two days, so divided by two. So if I ship 1,500 liters a day at 4 percent butterfat, I would use up my 60 kg of quota." He pulled out his phone to make a more detailed calculation. "So now watch . . . let's say our butterfat goes to 4.2." He types in 3000 (liters) × 0.042 (% butterfat) ÷ 2 (days). "Now I'm filling 63 kg of quota, rather than 60 . . . now I'm over quota. So how many liters of milk do I need to ship every two days?" He did the reverse calculations. "Two thousand nine hundred liters and now we're back to 60 kg. So I can ship 100 less liters a day if I get my butterfat up by 0.2%. That's how it works."

In practice, this means farmers can fill their quota with less milk volume—and therefore fewer resources—by increasing the butterfat content. One farmer I interviewed recently had to sell several cows because his butterfat was so high he was constantly producing over his quota.

And what affects the butterfat percentage? Genetics, health, farm management, and, critically, feed. As one farmer mentioned, "A lot of studies have suggested the high-yielding farms are more efficient at making better use of the feed." And indeed, this is right. The type of feed and its processing (dry vs. wet) is critical for dairy farm efficiency, milk production, and cow health.[19] Each of the farmers I spoke with focused particularly on alfalfa. With its high protein, calcium, and fiber, alfalfa "is like gold" to dairy farmers, and most dairy farmers in North America would readily grow it if they could, as it results in "lower feed bills and higher milk production."[20]

Farmers can usually buy "dry" alfalfa with ease. The alternative, fermented or "wet" alfalfa, preserves more nutrients, aids in digestion, and converts energy more efficiently into milk production. But the process of fermenting alfalfa is much more difficult. Alfalfa must be cut at the right time and chopped to the correct fineness to maintain proper moisture, then quickly packed into a bunk at a specified density to reduce nutrient losses and enable proper fermentation.

Harvesting, fermenting, and storing alfalfa silage is thus logistically (and biologically) complex; most farmers I spoke with do not trust this process to others, nor were they confident about the quality of silage available for sale on the market. After all, southwestern Ontario has some of the highest-quality land and grows some of the highest-quality alfalfa. Why not produce it oneself? Lena summarized the economics of feed well, so I quote her at length:

The problem with buying it is you don't know the quality you're getting. The benefit of supplying yourself is you know exactly what the quality is. And, you can imagine, what you put into the cows affects the output. If you put shit feed into them, you're not going to get good milk out of them. That's how we make money. We need to make sure we have a high-quality feed. So, the high-quality stuff is super expensive [on the market]. Usually, the stuff that gets sold, it can come from any which way, the US or Europe; you don't know if the crop got rained on. That's something we take pride in on our farm. We get good-quality crops. And that's a challenge every year, but you know when you get a good crop in, it supplies about three months' worth of food for cows, you need good stuff. It really affects your production. You need good stuff.

The farmers I spoke with would rather not rely on the whims of the market when they have the land and equipment to produce it themselves. A year with limited precipitation, said one farmer, could spike hay prices and suddenly make a worm cash crop not so lucrative. Even Harrison would never devote the land he *needs* for alfalfa feed to worm picking. "There's a lot of value to having your own [alfalfa]. You can always buy in a tractor-trailer of grain corn or something like that, but to me, we always have to grow our [alfalfa]." Bradley told me the best hay you can grow is your own, and if you're short one year, "other farmers won't sell good hay."

Nor do farmers enjoy paying for something they can make themselves; for most of the farmers I spoke with, the whole point of farming is to be self-reliant. Darla, the farmer who has absolutely no interest in pickers, was clear: "That's the whole objective of the farm, to contain ourselves as much as possible. So we have managed our nutrient supply over the years, always managed well, stored them well. So we don't buy commercial fertilizer. The whole idea is to close our loop so we're not subject to the vagaries of the marketplace." Daniel was perhaps most direct when asked about the economics of buying versus growing feed: "I don't sell alfalfa; I sell milk." Alfalfa is rarely an exchangeable crop, but is a cost of milk production that should not be commodified but produced directly by the farmer.

Coleman was the only farmer I spoke with who sees worm picking as an economic boon to his business, rather than a bonus, and I asked him if other farmers may overemphasize self-reliance too much—each owning millions of dollars worth of their own tractors, combines, harvesters, spreaders, and drag lines as a means of independently securing their high-quality feed. His eyes lit up, like it's something he'd been wanting to discuss for a long time: "I'm afraid a lot of farmers' independence skews their view on economic efficiency. And most times, the cost to own, harvest, and process is so much more than the value of the feed." He told me he had thought through the "economics of hay." "You know what a bale of hay is worth—a 500-pound bale of hay?" he asked me. "Twenty-nine dollars. That's the cost of harvesting nothing."

Coleman offered a different take on the alfalfa market. He has done his research on the alfalfa sellers. Whether it's the more expensive wet feed or the cheaper, less nutrient-rich dry stuff, "when I buy it, it's exactly what it is, and it's delivered to my door when I need it." He bypasses the required investments in storage facilities, and avoids the hassles of harvesting and controlling fermentation in a bunk. He has no risk to the harvest with weeds, pests, unexpected precipitation, or a drought. "And I can pick and choose what I buy, and get it analyzed and know exactly what I'm buying." I raised issues that other farmers had brought up, and he was not oblivious to the vicissitudes of the feed market. "So far it hasn't been an issue. You have to plan ahead." He only wished he had more land   not to grow crops but rather to rent it to the worm pickers. "Farmers are so fiercely independent, and they do things in the name of independence, but not necessarily economic common sense."

TO THE VICTOR GO THE SPOILS

Any southwestern Ontario dairy farmer growing their own alfalfa is de facto growing a lot of worms. Some farmers will take advantage of the last cut arrangement and profit an easy $200 to $300 an acre. Farmers like Scotty and Christopher will judge their feed stores after the first or second cut of alfalfa and decide whether to rent their field in July or August for a few hundred dollars more. These farmers do not engage in full-year contracts, however, because of the premium they place on self-provisioning their feed. So who are the big worm rentiers—the ones who command rents of $1,000, $1,200, or $1,500 per acre and forgo planting crops for an entire year or longer? In every case save Coleman, the farmers that rent their fields to pickers for the whole year have more land than is necessary to grow feed for their dairy herd. It is rare to be in this fortunate position. Most farmers I spoke with were typically short on land, both to grow enough alfalfa and to spread manure according to nutrient management plans. For farmers with *more* land than they need, however, the options are simplified. They cannot buy more cows and increase their milk production, as their US counterparts could. Instead, they could either grow more cash crops or rent land to a neighboring farm. Or they could take the bag of cash from the worm pickers. All this extra money also comes on top of making a living with guaranteed prices through Canada's supply management system, a hotly debated topic not only between politicians trying to maintain farmer support, free market proponents, and people who want cheaper milk but also between the protected and secure dairy farmers and the cash-cropping grain and oilseed farmers who find themselves far more exposed to unconstrained market forces. Follow the discussion boards or comments sections of online newspaper articles about supply management and you will quickly discover that "farmers" are not a homogenous group. The economic realities are distinctly different for cash-croppers, and their complaints go beyond the "unfair" government support that dairy farmers receive. Instead, they argue that supply-managed dairy farming actually undercuts cash-croppers by inflating land prices. Worm picking—for those cash-croppers who are aware of its value—is like the proverbial cherry on top of an intricately protected ice cream sundae that they will never get to eat. "There's a lot of animosity toward dairy farmers because they're in a controlled industry and they are already profitable," Coleman told me. "So the wealthiest, perceived as one of the wealthiest groups of farmers in Ontario can benefit from another side benefit."

And the benefits do not stop there. On top of the increased cash, several of these farmers renting their fields to pickers for a whole year claim yet another surprising benefit—one that initially runs counter to everything we've learned about worms. Renting land to worm pickers does not decrease soil fertility but, as two farmers told me, actually improves it.

. . .

Walter was not interested in talking about his worms while sitting at a table; he wanted to show me. I got in his new pickup truck to drive along the narrow pathways to his worm-picking field. Like all the other worm-picking sites I had seen, it just looked like an empty field. He also had a heap of dry manure he was spreading out to help fatten the worms. Then he said something that surprised me: "I'll be honest with you. The best thing that's ever happened to this farm is getting worm pickers. Over the past years, my land has been getting better and better and better." In the recording of the my interview, I responded with a simple, but obviously unconvinced, "O-*kay* . . ." From the transcript:

ME: You say it's the best thing that happened to you. Why is it the best thing that happened to you?

WALTER: Because I get better crops. I think it's a tremendous idea.

ME: You get better crops *because* of the worm picking?

WALTER: Yup. Oh yeah.

Soil research clearly demonstrates the mechanisms by which earthworms bestow benefits to the farmer—their contracting cylindrical bodies opening pores in the soil, the calciferous glands coating burrow walls with microbiota, the digestive tracts mixing soils and providing nutrients for soil biota and fungi. Such knowledge comes through experimental research that isolates and controls variables to identify how the physical, chemical, and/or biological mechanisms operate.[21] This research process must bracket out confounding factors to isolate cause and effect. At the same time, the research we choose to do and the variables *we* decide to study are social processes. Much of the research around soils (and earthworms) is done not for the sake of the soil itself but rather in service of industrialized agriculture for the explicit purpose of increasing crop yields.[22] This creates a circular loop of knowledge creation and application. Industrial agriculture influences soil science, which influences farmer decisions, which influence science again.

When we look at the "real world," however, confounding factors cannot be controlled, and previously well-established mechanisms may not explain the driving force in environmental transformation. Walter suggested that taking earthworms out of the soil got him better crops. This is largely the opposite of what research has demonstrated. He told me his neighbors think he's crazy, and said his father would certainly not approve if he were still alive today. To unpack this ecological puzzle, we need to move beyond isolated and controlled experimentation and relate scientific knowledge to broader socio-ecological relations. Looking at it from this angle, the puzzle of removing earthworms and getting better yields becomes very simple and obvious, like when a magician reveals the mechanism behind their seemingly impossible trick.

When I spoke with Walter, I still couldn't figure out the trick; so he said he would show me. We got back into the truck, and he posed a question I hadn't been asked in several decades: "Do you read the Bible?" I nervously laughed and responded in the past tense—"I have." He carried on, "If you let the land rest for a year—remember in the old times, in Israel. . . . It does amazing afterwards." He was likely referring to Leviticus 25: "And six years thou shalt sow thy land, and shalt gather in the fruits thereof: But the seventh

FIGURE 6. Walter's cornfield after a year of worm picking. Photo courtesy of author.

year thou shalt let it rest and lie still." Much like the Sabbath is a day of rest and renewal for Israelites, the seventh year is a time for rest and renewal of their land.

We parked next to a cornfield. Walter told me to stand in the field and he would take my picture so I would have photographic evidence of the superior corn that comes after a year of worm picking. "Have you ever seen corn this high? It's over 10 feet.... Hold your hands up." I raised my arms and he snapped the picture.

The Leviticus trick is simple. Walter doesn't cultivate land during the worm-picking seasons. The tractors and combines are not compacting soil, nor are nutrients being sucked out of the ground and into the crops. He is getting better yields, it would seem, because he is being paid exorbitant amounts of money to allow his soil to "sit and lie rest" for an entire year. Not only does integrating worm picking into his farm management enable a fallow year, but he continues to spread manure on the land, which is quickly gobbled up by the growing nightcrawlers who rapidly digest and expel castings throughout the soil. "No one knows the value of the casting, which I think is extremely high, and it will be my best field of corn." He noted a symbiotic relation. "Put it this way: if you take care of the worms, the worms will take care of you."

Walter has connected how soil and earthworm dynamics are not isolated "things" but interact through an assemblage of farming practices that are embedded in the political economy of Ontario dairy farming. Not all farmers would receive the same benefits if they rented their land to pickers. As I suggested earlier, worm picking could very well negatively impact soil dynamics in organic production systems—not because of the specific mechanisms by which earthworms impact soil dynamics but because of the broader socially constructed organic agricultural practices and standards that regulate tillage practices and manure application.

Coleman was the other farmer who noticed his improved yields. He was the most direct in response to my question about why he has worm pickers. "My response is twofold. Yes, it is economically beneficial, but what it does for the land is unbelievable." Again, I was slightly taken aback by his surety. He too noted—albeit without the biblical references—the immense benefits of leaving a field fallow. "There is no compaction and you're not taking anything [like nutrients] out." Though he has not tilled his land for twelve years, even planting and harvesting the nitrogen-fixing perennial alfalfa still comes at a financial and ecological cost. He went into more detail than I could understand about the gargantuan equipment that characterizes contemporary large dairy farms and the impact of its sheer weight on the soil. "You have to do three cuts, so you have three times the harvest cost, some Dutch people do four or five. [They cut] every thirty days—really hard on the land."

Coleman also saw the benefits going beyond a rest from compaction. He described integrating worm picking into his farm management as a "perfect conservation cycle." The alfalfa plant fixes nitrogen, holds soil together, and minimizes erosion from wind and rain. Worm densities rise and their bodies grow larger as they are fed with straw-based manure while "casting into stable forms of nitrogen, immediately accessible to the plant." Meanwhile, picking the large worms frees up resources for juveniles, which in turn kick-starts reproduction, leading to new pore spaces and aggregating soil at deeper levels—critically important for his clayey soils. Like Walter, he sees a noticeable difference in yields. "We don't buy fertilizer for two years after a worm-picker field. We'll have 200 bushels of corn per acre after that."

His reason for having the worm pickers is thus the complete opposite of what farmers like Geoff, Darla, or Peterson believe. "The opportunity to build soil's organic and nutrient level naturally [though year-long worm picking] is worth more than all the financial compensation we are getting now." At the same time, with limited land, he does negotiate rental prices. He got a decent $850 per acre in 2019. "What's the minimum you would take?" I asked. At first, he was hesitant to say, not wanting to risk his negotiating leverage. "Just because I'm so tight, that I have to buy feed and rent land. It's gotta be $600." Other farmers with more land set their base amount at $500. Either way, it seems that the nightcrawler capitalists are paying more than they need to.

CONCLUSION

Rent relations emerged in the nightcrawler industry because of the capitalist's inability to control and design the production of *L. terrestris*. The nightcrawler capitalists cannot accumulate according to their own preferences. They must move into and around the material realities that confront production, distribution, and consumption. The systemic imperative to accumulate capital means that they must access specific socio-ecological conditions that are more productive than those of their competitors. Thus, far from being a passive victim of their own commodification, nightcrawlers are actively co-constituting the bait worm industry in their relation to gray-brown luvisols, cow manure, alfalfa, no-till farming, and Canada's supply management system. The worms structure the conditions of possibility for accumulation and, through historical contingencies, the opportunity for rent.

The dairy farmers are accidental rentiers whose confined feeding operations have inadvertently turned their alfalfa fields into the most productive ecological conditions for nightcrawler production. However, trying to get them to play the role of rentier has been difficult. Nightcrawler capitalists are dependent on the whims of the worms and the rent demands of farmers to continue producing surplus value. How is worm picking still a profitable venture under such conditions? How do nightcrawler capitalists organize their rented land, capital, and labor in a competitive industry that can only formally subsume nature? And perhaps most mysterious of all, why is the industry underground?

_____

FOUR

Underground Capital

It's pretty complicated how you pick these things, because they
are alive, and you have to make sure they stay alive. Otherwise,
you lose a lot of money.

APOLLO BAITS, cooler

And it all happens in the middle of the night out in the country.
There's nobody watching.

KURT, dairy farmer

I COLLECTED CONTACT INFORMATION for crew chiefs and coolers
however I could find it. Aside from a few prominent coolers, searching online
was the least effective method. Instead, I found phone numbers on worm
boxes, vending machines, and job advertisements seeking "worm harvesters."
Farmers also gave me phone numbers, but they warned me that the person's
English was not very good, or that their number might no longer be in service. Coleman told me with a smile, "A company we deal with is called [name],
and I'm sure he wouldn't answer your phone calls." But he was wrong. I called
the number and a man answered the phone. After I told the man about my
research, however, he hung up. This was not a rare occurrence. Numbers were
often out of service or went unanswered. No one ever returned a call after I
left voice messages. One time I spotted a worm truck (see below) parked at a
home near a commercial shopping center. I knocked on the door and
explained myself to the older couple who answered the door. Before I could
finish, the man behind shook his head. "We were out picking all night," he
said. "We're not interested in talking." And the door was shut.

"Yeah, a lot of walls go up in the worm business," DCA Baits told me. I
asked why phone numbers I had were left unanswered or out of service. "You
tell me," he said, suggesting I already knew the answer. Why is the nightcrawler industry so tight-lipped? Why does no one want to talk about the
buying and selling of this seemingly inconsequential lively commodity? At
the risk of making a pithy tautological statement, the nightcrawler industry

is underground because it is underground. There is nothing illegal about the buying and selling of bait worms, picking them from rented fields, nor getting paid piece-rate wages. The nightcrawler industry does not *need* to be underground, but much of it is. This merits an explanation.

What do I mean by an underground industry? The underground or informal economy is typically defined as the "sum total of income-earning activities with the exclusion of those that involve contractual and legally regulated employment."[1] Initial research focused on the Global South, where informal and underground economies flourished; they were characterized by low barriers to entry, familial ownership, small-scale production, high labor intensity with little technology, and unregulated markets.[2] Such informal economies are often structurally heterogeneous, involving short-term contracts, putting-out systems of production, subcontracting, and, among others, domestic wage work.[3]

One significant advantage that underground industries have in relation to the formal economy is how they can operate at the margins of legality—often with one foot in both legal and illegal activities. Sidestepping taxes and other institutional regulations gives such enterprises a competitive advantage.[4] With low barriers to entry and cash payments, contracts need not be enforced by state laws, contributions to welfare programs need not be paid, and income taxes and sales taxes can be avoided. They are spared "from burdensome and costly regulations that can prematurely sink them or compromise their growth."[5] This creates an opportunity for individual capitalists to retain bigger slices of a surplus that would have otherwise been taken away through state regulations.

The concept of formal subsumption is useful for connecting concepts of "informalization" in the Global South to "precaritization" dynamics in the Global North, where the standard employment relationship associated with Fordist labor regimes started to fray in the 1970s. Many commentators have noted how the rise of neoliberalism and erosion of worker rights has increased the size and scale of informal economies in industrialized countries, noting production regimes and logics of accumulation that bear a striking resemblance to formal subsumption of labor as described by Marx in early British industrialization.[6] Sometimes the growth of these informal industries operates at different stages of production (subcontracting), or in different sectors, while other informal economies blur the distinction between "productive work" outside the home and "reproductive work" in the home.[7] For scholars, the concept of subsumption enables analysis of heterogeneous forms of production and a diversity of labor regimes that may appear anomalous but are still firmly embedded in capitalist relations.

Industries operating through the formal subsumption of labor and nature are not necessarily underground. I would argue, however, that the transition to the real subsumption of labor and nature may require fixed capital investments and increased exposure to state institutions and regulations, as well as institutional visibility. Marx hints at this when describing how the Factory Acts—an initial regulatory response to textile industries powered by steam or water—were unevenly applied in other heterogenous labor regimes; those who produced commodities without steam power were spared conforming to the regulation.[8] In many ways, increasing profit through the increased productivity of labor and/or nature creates more visible systems of material production that cannot easily escape institutional regulation. This does not mean that such industries cannot avoid tax obligations or engage in illegal activity (such as price fixing or tax havens), but the characteristics of such actions must take a form distinct from those of smaller businesses or self-employed people operating through the formal subsumption of nature and labor. Furthermore, the material characteristics of the nightcrawler commodity shape the opportunity for fixed capital formation and must rely on preexisting available land and labor pools to produce surplus value. Adhering to the physiological and ecological behavior of the nightcrawler creates opportunities for some nightcrawler capitalists to gain a competitive edge in ways that are not always legal.

"The greatest frustrations lie within the complexity of the industry," one of the larger coolers told me. For him, the industry is a logistical nightmare: trying to connect disparate parcels of land with available laborers, only under appropriate weather conditions, all within a ferociously competitive business environment with few barriers to entry, cash payments, and the opportunity to evade taxes. With a little bit of capital, worm pickers become crew chiefs, crew chiefs become coolers, and coolers feverishly compete over a limited number of customers. Add to this the accusations of money laundering and biker gangs, and you can begin to see why I would be hesitant to pursue a career in the bait trade. On the other hand, if you want to try and make some money in the nightcrawler business, this chapter can show you how.

HOW TO BECOME A NIGHTCRAWLER CAPITALIST

Gordon emigrated from Eastern Europe in the 1970s and at first made a good living as a general machinist. But as the early-1980s recession took hold, machine shops started to close their doors. He soon lost his job, as did his wife. The timing was especially terrible, as he had an infant child and a 19.6 percent mortgage on his home. He looked everywhere in the region, desperate to find work: "I had to do something to pay the mortgage and survive." A neighbor down the road, someone in a similar situation, came up with a novel idea. He told Gordon, "Try to get . . . a used vehicle like a van or something. I got a couple of people, and my wife and your wife . . . and we are going to start selling worms." "I didn't know anything about worms. Absolutely nothing," Gordon told me. He had heard golf courses were good for picking, but the greenskeepers kept chasing him off. He went to city parks and boulevards, even around hotel parking lots, before securing permission to pick on a large orchard that was exceptionally productive. Over the next year his crew picked worms and sold them to the coolers in Toronto, until he was asked a straightforward question by, of all people, his wife's hairdresser: "Why aren't you selling to the [United] States?" He had never really thought about it. After all, he couldn't speak English very well and didn't know how to attract customers. The loquacious hairdresser, however, could speak quite well; she would go with him to the United States and do all the talking. Gordon's wife was not thrilled about having her husband go on a cross-border road trip with her hairdresser, but she allowed it under the condition they sleep in separate beds. The hairdresser went into every little bait-and-tackle shop they could find in the United States and would just "start talking to those people." Gordon would wait outside, smoking cigarettes. "Then she comes back and says, 'Good, I got two people.'" In time, Gordon sold his own house for a property with a large garage and bought a 24-foot refrigerated box truck. "That was my first cooler." When the economy started to revive, he expanded his business, ditching his small cube truck for a single-axis refrigerated transport truck, which slowly turned into three more.

Gordon's cooler origin story is similar to those of other coolers who entered the business in the 1970s and 1980s. The early worm barons like the Brennan Brothers, Bob Conroy, and Bob Druin all appear to have handsomely profited during the immediate postwar fishing boom, but the low barriers to entry in the sector allowed some recent immigrants to quickly move up the value chain with relative ease, endlessly frustrating the established coolers who were constantly losing customers and laborers to these entrepreneurial upstarts. One longtime cooler listed for me all his former employees who are now his contemporary competitors: "[Name's] father worked for me. [Name's] father worked for me. [Name] worked for me."

Nico was one of those former employees. He came to Canada in 1972 and picked worms in the summer while working as a janitor in the winter. But the following summer, he told me, "I go to help somebody to load, and, uh, by accident, I don't know, I see the invoice. They pay me $7 to $9, and they got $20." He realized the investment necessary to make these superior returns was minimal. "I decided, in 1974, to buy a vehicle. I bought a Chrysler van and started a business. Me, I drove my wife, my sister-in-law, my brother, and another brother who came from Greece. Like a family. Nine or ten people in the van."

As these stories suggest, one of the first things one needs to be a night-crawler capitalist is a vehicle to transport worms and people between fields, warehouses, and homes. These need not be large cube trucks or customized worm trucks to get started. Particularly in the late 1980s and 1990s, pickers would show up at cooling warehouses in what were referred to as "small cars," with a picker, usually a spouse, and some friends or relatives, who were picking worms and avoiding the middlemen crew chiefs. The son of one cooler I spoke with remembered as a kid waking up in the morning and seeing a long line of these small cars, one after another, waiting to sell worms to his father, a visual that I found similar to the final scene of the movie *Field of Dreams*—if the father-son relationship were solidified through worms instead of baseball.

However, the "small car" arrangement is no longer prominent. One reason is that the total quantity of worms from these "small cars" is much smaller than can be acquired from a larger operation, creating more time spent on administration time and transaction costs. It was also more difficult for the coolers to control for quality. The coolers I spoke with tended to have some mechanism to monitor the quality of the worms by tracking the fields, crew chiefs, and sometimes even the pickers. Are they big enough? Thick enough? Were they "pinched"? Are they overheated? One picker I spoke with also suggested there might be a more coordinated effort by coolers and crew chiefs to minimize sales by "small cars" pickers to prevent more independent producers from taking away both sales and labor.

From the 1960s to the 1990s, crew chiefs used whatever vehicle could carry the greatest number of pickers they could find. These were often cube trucks, cargo vans, and sometimes repurposed school buses, makeshift vehicles that adhered to very little regulation. "Windows" were simply cut out of the side of the box wall, and old couches were provided for seating. Car accidents also tended to happen in the morning, as the crew chief might have slept only

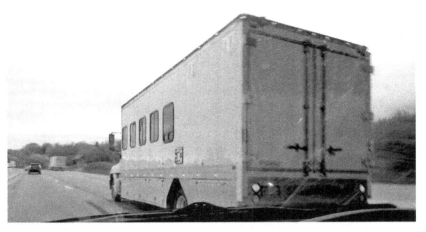

FIGURE 7. A worm truck. Photo by author.

intermittently or not at all. Persistent accidents involving worm pickers were cited as one reason to update the Ontario Highway Act.[9] Contemporary worm-picking trucks are custom-built according to the intricacies of the worm-picking industry. While at first sight, these trucks appear much like any other 20-foot box truck rolling along Highway 401 through Toronto, some peculiar characteristics suggest the vehicle is neither a transport truck nor a passenger van. Worm trucks have a series of windows along the front half of the truck, along with a folding door like a school or city bus. The front portion of the box is reserved for bench seating. The last part of the box is partitioned from the seating area and reserved for the stacked flats of worms. Air vents are visible from either the top or bottom (or inserted into the back doors) to circulate air for the compactly stacked worms, which must survive in the truck for hours. Undercarriage storage compartments also provide space to house worm-picking equipment and any other personal effects. The exterior of the box is also typically framed with small lights that increase visibility while it is parked at the side of narrow rural roads for most of the night.

To step up in the value chain, some smaller crew chiefs also invested in refrigerated containers, allowing them to store the worms for longer times and to sell directly to retailers in the United States instead of going through the coolers in Toronto. Instead of relying on his wife's hairdresser, Nico enrolled a cab-driving friend who "spoke good English." He had some client contacts ("I stole some numbers from the guy whom I sold the worms to") and told the friend to go to the States and find more customers. Bit by bit, his customer base expanded, aided by the continuing increase in the number of

FIGURE 8. Five hundred worms in a cooler's flat. Photo by author.

recreational fishermen during these decades. Even today, coolers can start this way. One recent picker-turned-cooler "learned the business quite fast" after a sneak peek at the crew chief's invoice. Their necessary capital investment? "A pickup truck, a trailer, and a farmer. That's how we started."

The next step up the value ladder is to invest in a worm warehouse, which can range from 400 square meters to over 3,000 square meters, the latter capable of holding between 30 and 40 million worms at any one time. When the worms arrive at the docking bays (or drop-down garage door), they are unloaded from the picking truck and dumped into the cooler's flats, filled with peat moss (or other bedding mixtures that coolers prefer to keep secret). The flats are then moved to a refrigerated cooler and stacked diagonally on pallets to ensure sufficient air for the worms. Lights are kept on to discourage the worms from wandering out of the flats.

Coolers will need additional space in the warehouse if they are prepackaging worms into cups of one or two dozen. Flats of worms are dumped onto tables and counted by hand by waged employees. Excess bedding, castings, and dead and dying worms are briskly swiped aside into a bucket below. It is at this stage when worms are sorted for the last time, mostly for length and health. The workers are also looking for "pinched worms." In these cases, the worm picker's fingers gripped the worm slightly too hard, and later the worm

FIGURE 9. A pallet of worms identified by farm. Photo by author.

FIGURE 10. Sorting and counting worms. Photo by author.

has turned a grayish color and is extremely lethargic. Indeed, it is quite easy to see, even for my untrained eyes. Some flats are filled with uniformly large nightcrawlers with a healthy reddish hue—signs of a good farm with lots of manure and a good picker. Even the workers can be astounded by their size, with one young woman deciding to keep a 19-gram worm as a pet. She named it Chára, after the 6′9″ NHL hockey defenseman Zdeno Chára. Flipping over another flat from a different farm or crew chief might show more variation: pinched worms, smaller worms, thinner worms, less lively worms.

The level of automation inside the warehouse varies according to the cooler. Some employees simply stand around a series of tables, sorting through the worms and putting them in cups. Larger operations will use a conveyer belt where worms are hand-sorted, then placed in cups that are mechanically filled with peat moss, capped, and then stacked on a skid ready for shipment. Most of the employees who prepackage the worms are women, working for an hourly wage on a seasonal basis. Larger coolers may employ between fifteen and twenty-five seasonal employees. Employees are often former worm pickers who have opted to forgo the potentially profitable nighttime labor in exchange for more stable daytime hourly wages. In comparison with the worm pickers, employees working in the warehouse are paid by check with necessary income taxes and unemployment insurance premi-

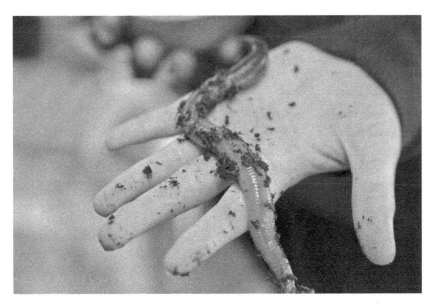

FIGURE 11. An earthworm named Chára. Photo by author.

ums deducted. This is important, as both worm pickers who are paid by check and warehouse workers can benefit from unemployment insurance, worth up to 55 percent of one's salary during the off-season.

In other warehouses, the employed women are a group of friends, perhaps neighbors, now with kids in school and free time on their hands. While they work quickly and efficiently, they chat about their families, national politics, or any other noteworthy community news. They say they are not doing it for the money per se (though they are paid); instead, as one woman told me, it gives them a chance to get out of the house and socialize. This relaxed atmosphere, however, is certainly not widespread across the industry. Some coolers accused others of underpaying their warehouse employees or keeping their wages off the books and sidestepping regulations such as overtime pay. Another cooler was adamant that I could not see their sorting and packaging process. The speed of packaging, the cooler felt, was their competitive edge; they confidently stated, "no one can pack faster than us."

All the coolers I spoke with owned at least one transport truck ready to drive across the border to ensure the product arrived alive and on time. Outsourcing transportation is risky for a commodity whose value depends on it staying alive. Parga Live Bait explained it like this: "If you rent a trucking company, it costs more money. And they don't guarantee it. The worms must get there right away. You must sell the worms today. And the company says, 'I can't go that way today. I'll go there in three days, four days.' But we need the worms there today! That's the difference. You have to have your own driver."

In this secretive business, one way that coolers can estimate their competitors' sales is by observing the value of their trucks. Apollo Baits notes one of his competitors has "a brand-new tandem truck like ours [that is, with double rear axles]. To buy a $200,000 truck for your own deliveries. . . . You don't buy it if you're delivering 20 million worms. They're probably between 50 and 100 million [worms]." Smaller operations might have a small fleet of sprinter vans for short trips within Ontario and smaller orders across the border.

(UN)CONTROLLABLE ECOLOGICAL CONDITIONS
OF PRODUCTION

Buying trucks, property, and compressors or renting warehouse space is simple. These are expenses like in any other business. You need a conveyor belt or a compressor? Get a loan, or cobble enough money together and buy it. You need raw materials or new technologies? Buy them on the market and put them into your production process. But these materials are only for the sorting and storing of worms. The machinery for actually producing the nightcrawler commodity—a manure-soaked, no-till hay field in southwestern Ontario—is not as easy to obtain. As discussed in the last chapter, dairy farms have accidentally become the most productive worm fields because of their no-till practices and massive manure application rates. But which farms and where? An alfalfa field near Woodstock has high worm densities, as do farms around Mount Forest, as well as London. How does a nightcrawler capitalist decide which land to rent? Two contradictory criteria must be balanced for selecting worm-picking land: proximity to labor and geographic diversity.

Crew chiefs and coolers are incentivized to rent farms as close as possible to the labor supply, the majority of which comes out of Toronto. The closer the productive field is to the city, the less travel time for both humans and the tightly packed worms. This makes Highway 401—one of the busiest roads in North America—the main transportation artery of the bait industry circulatory system. It cuts right through prime picking land between Toronto and Windsor, with smaller highways such as 8, 6, 403, 402, and 410 acting as capillaries that deliver pickers to the targeted rural land. In earlier decades, crew chiefs would pick up and drop off each picker at their homes in the evening and the morning, respectively. This ensured that recent immigrants without a vehicle could get to the fields, and the crew chiefs would have their labor, but it significantly added to the travel time—time for which the piece-rate pickers were never going to be compensated. Add to that Toronto's horrendous commuter traffic and pickers could spend upwards of six to seven hours on the truck during the "workday" without picking a worm or making a cent. Prearranged pickup points can significantly trim commuting time. But again, the location of the field will inform the picker's decision about their employer. Closer fields and more productive fields will attract more laborers.

The other option, becoming more common, is for the pickers to drive to the fields directly. Not only does this eliminate the travel time of picking up individuals, but it also allows the pickers to arrive at and depart from the field at their convenience. This is especially advantageous when they feel the number of worms on the surface does not justify their time or when they want to avoid commuter traffic. The crew chief will gladly accept any labor that shows up on their fields, and in fact, pickers with their own vehicles can hop between fields on the same night. The plurality of transportation arrangements ensures that any person willing to pick can get to fields to pick as many worms as possible. As one cooler told me, "There's a little method to the madness."

Several coolers—due to both circumstance and strategy—are based outside of the Toronto region. This shifts the circulatory transport system to access the same fields while using different roads and commuting times. "Time is of the essence for the pickers," said one cooler. Having a cooler in Kitchener-Waterloo—about a ninety-minute drive west of Toronto—permits them to employ pickers from smaller cities and rent land in much closer proximity. The coolers and crew chiefs in the Kitchener-Waterloo, Stratford, and London areas have a bit more flexibility and can rent fields in each direction while maintaining shorter commutes.

Therefore, on the one hand, the decision to rent is simple: find the most productive land at the closest distance to available labor. This attracts pickers and minimizes commuting time. But focusing only on proximity raises a different problem: nightcrawlers only surface under certain weather conditions. For an optimal night of worm picking, temperatures, soil moisture, moonlight, atmospheric pressure, and wind speed must harmoniously align, and this does not occur evenly across southwestern Ontario. A worm-picking field outside London, for example, may benefit from rain that never arrives in Mount Forest, or vice versa. Renting a couple of hundred acres just outside

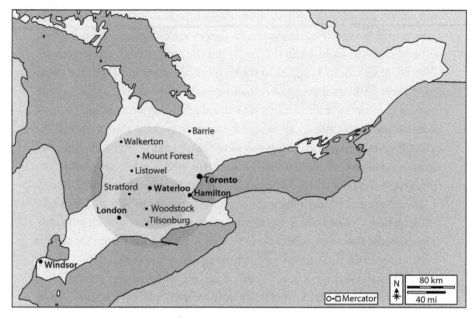

MAP 1. Prime worm-picking area in southwestern Ontario.

of Toronto might be wonderful for commuting times, but it restricts the number of nights when weather conditions align. Geographic diversity of rental land is essential.

Nightcrawlers can be picky and unpredictable about when they rise to the surface. To begin, the nighttime temperature must be below 17°C and above freezing (although temperatures above 10°C allow more variation in other weather conditions, such as wind speed). Coolers are adamant the worms should *not* be picked at temperatures above 17°C. DCA Baits told me, "Worms will probably still come up. But when you put them all together, they get all slimy and almost suffocate themselves in their bags." Ben, a large cooler, told me he watches the nighttime temperature closely each summer night to see when or if it dips below 17°C. When a crew chief shows up at his door after a hot summer night and the worms appear lethargic, he knows what happened: "You didn't wait [for the temperature to drop]! You picked them at midnight!"

Moisture is also critical. Many of us know this from witnessing an abundance of *L. terrestris* in gardens and on lawns and sidewalks after a night of rain. Light drizzles appear to be the most optimal, as the rain droplets do not strike the ground too hard, and the clouds block the moonlight. The crew chiefs, in particular, watch the rain closely: "It's always a daily question where you're going to pick worms today. Some days it's easy, if it rains everywhere, or if it only rains in one spot. They're forever watching the weather maps," said DCA Baits.

While the temperature might be ideal, fast wind speed can derail a good picking night. One picker I interviewed opened his phone to show me how he tracks the weather daily. That night, wind speed was predicted to blow at 15 km per hour. "That's nothing," he said. Over 20 km per hour, however, and he will want to investigate a bit further. If the temperature is cool, the high wind speed will drive the worms underground, whereas it may not have the same result in warmer weather. As wind speed increases over 20 km/h, the pickers and crew chiefs are less and less inclined to make the commute. All of the pickers and coolers noted that worms are particularly sensitive to cooler northeastern winds.

In the 1970s and 1980s, coolers subscribed to satellite feeds to follow where and when weather conditions would align for an optimal picking night. Unsurprisingly, coolers and crew chiefs were also early adopters of cell phone technology. Car phones and early cell phones further allowed them to have real-time information from multiple drivers at dispersed sites across the province. Apollo Baits told me he had them all: the Motorola car phone, the brick cell phone, the Razr flip phone, the Nokias, the Blackberry, and finally the iPhone. "Oh, fuck, it's amazing now," he said, noting that all weather information can be set up with the touch of a button.

While one can monitor the weather conditions, deciding where to pick is still a subjective decision, and one that may rapidly change. Even when all the conditions are aligned, the worms' behaviors still might not make much sense to the pickers. Nico remembered a farm in the 1980s. He dropped off the pickers, who were complaining there were too few worms. He drove around for forty-five minutes, checking on other fields, and returned, to find a woman had pulled 4,500 worms in an hour. "I said, 'what's going on?'" He saw the headlights rapidly crisscrossing the field, with each picker raking their fingers along the ground, grabbing worms by the dozen. Fifteen minutes later, they were walking back to the truck. For completely unknown reasons, these lively commodities had retreated, and the pickers had to look for another field.

Therefore, renting land from a variety of different geographically located farmers is paramount, and also offers a clue as to why nightcrawler capitalists don't just buy the land outright for the sole purpose of producing nightcrawlers. As of 2020, the average cost of an acre of land in Oxford County—prime worm-picking land—is close to $40,000.[10] A measly worm-picking field of

20 acres (assuming 20 acres were even for sale!), therefore, would cost $800,000. In addition, the nightcrawler capitalists would have to haul manure onto the land to fatten the worms, which might not be cheap or even readily accessible.

Relatedly, nightcrawler capitalists do not want all their worm-picking land in one region. Having a contiguous 500-acre plot of fertile, worm-filled soil might be useful when weather conditions are good, but it becomes a serious liability when such expensive land receives less rain than other areas around Ontario. Apollo Baits explained to me how, in 2019, he rented 70 acres of land for $70,000 but never picked a single worm from it; other geographically dispersed land was more productive. Purchasing the same amount of land could have potentially cost him $2.8 million. He preferred to take a loss of $70,000 for the year, knowing he had other land producing worms. Again, the relative inability of capitalists to control the worms impedes fixed capital investments in land and forces the rent relation on the coolers and crew chiefs.

INVESTMENT DECISIONS

The low barriers to entry, minimal fixed and circulating capital, and diversity of geographic locations create a plurality of production organizations. From the 1950s to the 1970s, the large coolers tended to operate vertically integrated businesses. They paid rent to the golf courses and hired crew chiefs, either through salaries or commissions, who in turn directly paid the pickers. As demand for worms increased and more competition developed, coolers began to subcontract the organization of labor and rent to the crew chiefs—a trend that continues.

One larger cooler has two of his own worm-picking trucks, mostly to transport his own worm pickers to his fields. He then relies on five other independent crew chiefs who may or may not pick on his acres of rented land. DCA Baits has one crew chief who picks from their 300 or more acres and relies on independent drivers for the rest. Apollo has about 300 acres as well, with three drivers. Others have more informal agreements with crew chiefs based more on personal relationships. But when the weather keeps worms in the ground and prices skyrocket, these alliances and informal agreements quickly break down, and the crew chiefs (and beneath them, the pickers) have significant bargaining power. During the 2014 polar vortex, DCA Baits remembers, "that was crazy. Some of those worm guys shop to the highest bidder every day. Even the worm bosses [crew chiefs], it looks like they have a union out here. They basically all just rally together and say we want this much and if you don't pay, we're going somewhere else. They basically had us all by the short and curlies for the whole year on that one."

The low barriers to entry also mean there are many actors engaged in hybrid roles (crew chief/cooler, picker/crew chief, independent/contracted), each competing against one another for a valuable annelid resource in the midst of a shrinking market.

In this context, reinvesting all profit to expand by building a new cooler, or investing in a double-axle transport truck, is a risky venture when a dry spell will send worm prices skyrocketing, leaving unfilled customer orders and interest on debt accumulating. As one cooler mentioned, "That's why I never invested in it beyond the necessity." I quote his reasoning at length:

> If you take your money and put it into the business, you're going lose it. You put enough to get you to the level you need to be to stay competitive. Like, if you try and make these giant moves that you want to double or triple your business, your sales—the debt would outweigh that. I know a lot of companies that went bankrupt because of that, and they started up again, but they remain much smaller. They lost a lot of money on that kind of stuff. You can't do that in this business. You gotta be really, really careful. You have to take small steps because one customer moving could derail everything. It's not your traditional business that you build something, then you sell it, which people do. I realized very early on that you couldn't do that in this business. And whoever didn't realize that and dumped money into the business, they're not going to recover any of that.

Surplus profit is thus not immediately reinvested in expansion, research, or new technologies, but instead gets diverted into other ventures outside of the industry. One worm cooler owns considerable amounts of high-priced property in Toronto, having invested in real estate in the 1970s and 1980s. Others have done the same both in Toronto (when real estate was more affordable) and in the relatively cheaper Hamilton and Kitchener-Waterloo markets. Other coolers have invested in completely different businesses, such as car washes and other service industries. They recognize their worm businesses are not a store of value, nor are they capable of "cashing out" to a potential buyer. Once all the fixed capital is added up—the trucks, the sprinklers, the Bobcats, the compressors, and so on—as one cooler said, it might be worth a few hundred thousand dollars in the used industrial equipment market. A $20,000 compressor, the cooler said, will be worth $2,000 five years later and will cost more than that simply to dissemble it and remove it from the warehouse.

The variable capital—that is, the pickers—have few allegiances and do not come with the business, so to speak. "You say your business has employees? The pickers?" Apollo Baits asked rhetorically. "Well you do now, but when a competitor offers a dollar more per thousand, they will be gone." The same holds for the American and European companies purchasing the worms. Another younger cooler said he was initially going to purchase the business from an older cooler until he realized there wasn't much to buy: "They want to sell, but what am I buying? You can't buy customers; your customers are gone. Buy your equipment? I got my warehouse." Even for a big cooler, Apollo Baits suggested, "nobody would give them a penny for their business." I suggested that there might be value in having a stable clientele, a multitude of wholesalers and retailers ready to buy up hundreds of millions of worms. Apollo Baits laughed. "As soon as [name] dies, it will take a minute [to get his customers]. I already have my strategy.... We all do." He was frank: "The business has no value. None of our businesses have value."

Instead, coolers continue to profit by loosely maintaining the relationships they can. Each of the older coolers noted they still have their longtime customers from the 1970s and 1980s. Gordon noted, "[Name] is eighty-three years old. He's still healthy and alive. He was the first customer buying worms. And he's still buying from me. And I got [name], he's ninety-one. His son took over, but he's still buying worms from me. We are friends. I know his great-grandkids. I know his kids. I know his whole family." These relationships developed in the heyday of worm picking in the 1970s, and there was a sense of appreciation and acknowledgement of how US bait buyers enabled recent immigrants to make a life in Canada. "Those people, I like them," said Nico. "They gave me bread to eat for so many years. Some of those people, I don't want to lie to them, I don't want to let them down." He has several customers that do not even try to negotiate prices: "I have a guy—haven't spoken to him two to three months. He trusts me. He calls, 'I need this and this and this.' Okay, no price, no nothing. He trusts me and I trust him." Thus, the real value of a bait business is not in its fixed capital, the number of employees, or the customer base, but rather "your relationships with the pickers and the customers." The problem is, "you can't sell that value."

But what if there *were* a way to add value to the worm in a way that could be controlled? *L. terrestris* may be resistant to cultivation, but what if—once sitting in a cooler—it could be turbocharged to enhance its advantageous characteristics as a bait commodity? What if the worms were to become more visible to the fish searching for food in the deepest of lakes? Or what if they developed a scent that could rapidly diffuse through the water and reach the fish's olfactory pits from greater distances? What if the real subsumption of nightcrawlers was not about increasing their reproductive productivity or size, but rather increasing the efficacy of the bait—a super-worm with unique attributes capable of attracting the largest bass as no other nightcrawler could?

In 1993, Paul Giannaris wrote an academic paper titled "Oligoribonucleotides Containing 2',5'-Phosphodiester Linkages Exhibit Binding Selectivity for 3',5'-RNA over 3',5'-ssDNA," published in the *Nucleic Acids Research Journal*. This is one of twelve esoterically titled journal articles he coauthored while conducting his PhD research in chemistry at McGill University. When his father died at fifty-three, however, he left the university laboratories and withdrew from his PhD to take over the family business. Paul's father was a Greek immigrant, and shortly after arriving in Toronto in 1977, he purchased a van and headed to golf courses and pastures around Toronto to pick worms with his wife and cousins. Over the years, the business grew, and the father became a cooler selling up to 10 million worms a year during the 1980s. Paul took the reins of the business and expanded. For the past thirty years, he has consistently increased worm sales, and now ships worms around the world. He was quick to recognize, however, that there were few opportunities to increase profits through the production end; worms must be picked by hand, at night, under particular weather conditions, by people willing to squat in muddy soil over the course of eight hours. But what if he could increase the value of the nightcrawler after it was in his hands but before it was sold to his customers?

In 1999, Paul filed patent number US6240876B1, which claimed the "ability of a bait to attract fish is at least partially related to its color."[11] This, of course, is part of the advantage of artificial lures and baits, with their shiny spinning silver, fluorescent threads, and colorful rubber. The live earthworm has many superior attributes—its liveliness, scent, and taste—but it can never escape its rather dull, earthy hue. Fishermen have long known this and there had been previous attempts at "coloring their live worms." An earlier patent filing, for example, describes a colorant made of "ground cornmeal, a coloring agent and anise oil" that would be purchased by a fisherman and applied to the worm. A later patent found that worms could change color themselves with the addition of concentrated food coloring to earthworm bedding. Paul—the chemist-turned-cooler—experimented with these prior methods but was left unsatisfied. His patent states: "It has been found by the inventor . . . that many presently available edible food colorings are unsatisfactory for coloring live bait worms, and that the amount of food coloring recommended . . . is in many cases detrimental to the viability of the worms." The worms were dying, and the color-infused cells quickly faded in retail packaging. But the idea seemed right.

Over the next several years he conducted his own experiments with the methodological precision befitting a former doctoral candidate: "Each powdered colorant was mixed with peat loam medium and added to an 8 oz cup containing 12 healthy and natural worms. Concentrations given as mg colorant per 8 oz cup containing about 120 g peat loam. Worms were inspected for color and viability 24 hrs., 36 hrs., and when still viable, 3–4 weeks later."

After extensive testing, he made up his mind: "The most preferred colorant according to the invention is D&C Yellow No. 8, also known as uranine, which is the disodium salt of D&C Yellow No. 7, also known as fluorescein." Fluorescein, Paul writes in his patent, "provides the worms with an intense fluorescent chartreuse color." Chartreuse, for any reader without a paint swatch nearby, is a yellowy-green color named after the French liqueur developed by the eighteenth-century monks of Vauvert. Put another way, it makes the worms look radioactive. When I spoke with Paul in person, he further extolled the virtues of fluorescein. "That stuff's amazing. It just glows like crazy." And it had no discernible impact on the worms. "I tried to kill worms with it. I couldn't kill a worm."

As a live bait, the benefits of the fluorescein-fed chartreuse worms were immediate. "You go down thirty feet, and you can see this thing glow. It just picks up the light." While other coolers and fishermen I spoke with were not entirely convinced it actually attracts more fish, Paul assured me it was no gimmick: "We had fishermen going on tournaments doing it. We were developing the product and marketing the product." A tournament in Michigan found a fisherman hooking a 5 kg walleye with the green worm. His first press coverage came from an *Los Angeles Times* article that correctly stated: "These worms are made, not born," and they were selling fast. As one bait dealer in California said, "From the first time I saw them, I knew they were going to be popular."[12]

At the time, in the early 2000s, a dozen worms sold for between $0.80 and $1 US, and Paul was certain that "it could sustain a price increase," far beyond the minimal cost of the fluorescein. He called these enhanced nightcrawlers Nitro Worms and, with patent in hand, began selling them for $2.50 a dozen in the United States.

By soaking the peat moss in fluorescein and infusing the water-soluble dye into the cells of the worm, he was able to create relative surplus value through the real subsumption of nature. Relative surplus value, however, is ephemeral, as competitors started playing catch up. He had a patent—a source of rent—protected by law, but he knew it would be difficult to enforce: "A patent is only as good as your pocket is deep." He tried a different tack: he wouldn't market his Nitro Worms to his competitors' clients, but instead sell the other coolers his worms at wholesale cost. This arrangement, he suggested, would allow everyone to participate and benefit from the added value of the product. This worked for a short time and the price per dozen settled on $1.50 US, a 50 to 60 percent markup from the boring earthy-hued worms. It didn't take long, however, for "the coercive laws of competition" to come out in full force. His competitors combed through Paul's patent filing; they suddenly became interested in the alchemy of green worms and started producing the worms themselves. Paul urged against undercutting the market price, not because of the patent, nor even his personal profit—although both would be severely impaired—but because he knew where this would lead. "What they don't realize, the second they do that, the price isn't $1.50 anymore. It's 80 cents plus a penny." The relative surplus value erodes through competition until it provides no additional benefit to the instigator or the followers. "So I tried to explain it to these imbeciles [the other bait dealers], but they didn't understand."

And after a few more years, the price of the turbocharged Nitro Worms (and their green knockoffs) was driven down completely. Now, he said to me, "They're selling it for 5 cents more than the regular cup. Actually, they're not even selling it because no one is marketing it. No one is pushing it. No one knows about it. . . . We should have been able to provide that product to a good market. But everybody got greedy—standard thing—and then it got dissolved."

"Getting greedy" individualizes what is really a systemic contradiction. Intra-industry coordination could have set the value of the Nitro Worm far above the cost of production and allowed each of the coolers to secure additional profit through a form of absolute rent. But this never happened. I asked coolers why there is no worm-industry equivalent of the Grain Farmers of Ontario, the Ontario Tender Fruit Growers, or the Ontario Fruit and Vegetable Growers to coordinate industry-wide actions. There is no Association of Ontario Worm Growers, nor a Nightcrawler Marketing Board. The only advertisements I've found even promoting the nightcrawler commodity as bait come from a regional US wholesaler, DMF, the largest purchaser of Canadian nightcrawlers. They produced a series of commercials full of double-entendres, entitled "Bait like a Master," which contained such stimulating dialogue as "Oh, it's bigger than I expected! Does it always wiggle like this?" No such tongue-in-cheek commercials come from the Canadian coolers. The closest thing they have to a lobby group is the Ontario Live Bait Angling Association. Nightcrawlers are, after all, a living bait, and coolers have participated in the association in past. However, its primary mission is oriented toward *aquatic* live bait policies, which are becoming more restrictive in Ontario.

Few coolers coordinate with each other, and each of the coolers I spoke with helpfully reminded me not to trust their competitors. "We talk to each other, even if we're not telling the truth," said one. And this is a significant impediment to the overall profitability of the industry. Partnerships between coolers have existed, often between immediate or extended families, but they never seem to last more than a year or two. As soon as additional profit becomes possible for one party, the undercutting begins. Gordon wishes there were more coordination between coolers to control the flow of worms into the United States, much like a cartel. He told me there might be a few pockets in the United States still producing lower-quality worms, but there is no other place in the world outside of southwestern Ontario that can fulfill the demand. It's a monopoly, he said, that is squandered every year. Put more crudely, Gordon said, "if we are united, we are going to be able to fuck the Americans" by raising prices. "But we are stupid."

Instead, coolers consistently undercut prices and wages to increase market access and attract labor, sometimes only making money through the exchange rate with the US dollar. This drives competing coolers crazy, as their customers will either chase these below-cost worms or ask for discounts. "This person [a competing cooler], they got no brain!" Gordon told me. He doesn't understand how these producers are making money. "He says, 'I want to sell for volume.' Well, I want to sell for money." Gordon suspects he might have a similar net revenue by selling 60 million fewer worms. Newer coolers, or crew chiefs who are taking the next step of selling directly to customers, also sell below cost to develop their customer base. Every now and again, an email circulates across bait dealers in the US explaining there is a new supplier ready to offer the best deals. For coolers like Gordon, it's an unending barrage of offers that persistently drives prices to the break-even point and below. The past year, he had to raise his wholesale prices by the equivalent of $1.25 per dozen to cover the labor costs, shrinkage, and farm rentals. "But then Titan Bait shows up for 98 cents . . . deals for a year or two then raises the price." Gordon lost this customer, who then quit Titan Bait when the price rose, and went with another cooler, Johnny, for $1.10. "Johnny raised the price, they quit Johnny, and they take me, again at $1.25. And now Lake Ontario Bait shows up. Today, 93.75 cents, he's selling a dozen! Then this guy, Fred, he's selling 87 cents. And I can't sell it at that." It appears neither could "Fred," as I later learned. He couldn't repay his loans by selling so cheap, and quickly went out of business.

The cycle continues. New players enter quickly, make a grab for customers, depress prices, anger competitors, and then disappear. The coolers noted the structure of the business is not sustainable, with several suggesting there is a reckoning coming over the next decade: "The expenses keep going up. You try and sell more worms. That's what I'm saying, this business is going to consolidate, you can't survive on this. . . . My profit's gone down 50 percent in the last ten years, but I can live with that because I'm old enough and I've invested elsewhere. But somebody coming in, they're not going to be able to cut it. You need consolidation, but it won't consolidate; you're among idiots."

Nightcrawler capitalists do have another option to retain greater surplus value than their competitors—one that takes advantage of the cash transactions, high-volume sales, nighttime labor, informal contracts, low barriers to entry, and lack of regulatory oversight that characterize the industry. This option is also one of the reasons why it was so difficult for me to connect with crew chiefs and coolers, whose phone numbers were often out of service. As Apollo Baits explained: "There's a reason for that. It starts with T and ends with X."

TAX FRAUD AND MONEY LAUNDERING

One cooler suggested "probably 90 percent of this industry is in cash." Duffel bags of money permeate the industry all along the value chain: cash to pay the pickers, cash to pay the drivers, and cash to pay the farmers. Much of this underground activity rises to the surface during droughts, when coolers are scrambling to get their hands on the rare and high-priced soil fauna and capture the attention of journalists. "Worms are like gold this year," said Peter Kotsifas in a 1979 newspaper article.[13] One morning he entered his warehouse to find $42,000 worth of worms had simply vanished. There were also anonymous threatening phone calls, slashed tires, and poisoned worm bedding. One cooler found two Molotov cocktails under the gas tank of their truck. At the time, an anonymous cooler told the reporter the bait business "is like a syndicate; it's difficult to penetrate. I can't talk freely because I don't know who will read your paper. Some enchanted evening, I could get a two-by-four on my head."

In the 1980s, there was also significant conflict between picking crews over access to the sites; these were termed the "worm wars" by the *Globe and Mail*. Pickers got into violent altercations with each other, throwing punches and hitting each other with sticks and pipes. A police officer at the time recalls getting called to a golf course where two competing groups were throwing boxes of worms at each other. The policeman said: "We're out here in the middle of the night and it's wall-to-wall worms."[14] In another violent incident in the 1990s, "thirteen people were injured, some with steel pipes, while four cars were badly damaged, and a van was set on fire."[15]

Some of these newspaper articles caught the attention of the Canada Revenue Agency (CRA). I spoke with one former agent who dove into the intricacies of the industry between 1985 and 1997. "The guy knew the business in and out," one cooler remembered with a laugh. The agent tracked income and expenses, quickly figuring out what a million worms looked like and how much peat moss would be processed by the worms as they sat in the coolers. He also found notebooks from coolers and crew chiefs who tracked their under-the-table payments, connecting cash payments to matching Social Insurance Numbers (SINs), signatures, and bank deposits. He would show up at pickers' houses only to find many of the recorded SINs belonged to deceased people, children, and other individuals who clearly did not have the ability to pick worms. At one residence, he asked to see a man named Min, who had been paid thousands of dollars in cash by a cooler. But when Min appeared at the door, the agent knew someone else was using his SIN: "He was paid $5,000 in cash for picking worms, but didn't have any hands."

He eventually had a team of ten investigators and would visit the coolers and crew chiefs in the area every two years. "That's when they got to know me," the agent told me. He remembers searching a home and finding tens of thousands of dollars packed in a suit hanging up in a closet. In another case he found two semiautomatic weapons and $10,000 in cash under the front seat of a worm truck. Actors in the picking industry quickly came to recognize his face and would glare at him down the aisle of the local grocery store; in one case, someone threatened to poison him when he sat down at a restaurant owned by a former cooler he had recently put out of business.

The agent's investigations put the bait industry on the CRA's radar and, as a result, the industry started to formalize, at least at the cooler level. The actors in the industry, as one former crew chief told me, should not be thought of as working "legally" or "illegally," but instead should be viewed on a spectrum of legitimacy. When I asked a cooler about one of his crew chiefs, he suspected "he might be 80 percent legit." In contrast, another cooler was adamant that every one of his transactions was properly recorded—unlike his competitors who pay their farmers and pickers in untraceable cash. So when I spoke with one of his contracted farmers, I was mildly surprised that the farmer *also* received an untraceable duffel bag of cash in addition to the recorded checks. It is harder for coolers who export the majority of their sales to the United States or Europe to get away with various tax shenanigans. "If you're like me or [name] it's really hard to do. . . . All my stuff gets marked the second it leaves the country," said Apollo Baits. "The government knows I sold 2 million worms. . . . If I show I bought a lot more than I sold, then they'll walk into my cooler. 'Where's the other 2 million? Where are they?' Forensics are easy for warehouses because we export." So where does most tax fraud occur? "It's easy for the people before us," Apollo Baits said. "The drivers [crew chiefs] are really the issue."

How might one interested in tax fraud get away with it? There are two broad ways to avoid (or retain) taxes in the bait industry, although these techniques are easily deployable for other small businesses in Ontario willing to take the risk. In the simplest scheme, a crew chief sells worms to a cooler and charges the Harmonized Sales Tax (HST), which adds 13 percent to each transaction in Ontario. Then, instead of paying that tax back to the government, the crew chief simply holds on to the extra cash. It is obviously a risky strategy; there will be an invoice in the cooler's hands with evidence that HST was charged for the sale. This can be mitigated, however, either by not registering an HST number with the CRA (which will add an extra step for the tax authorities) or by persistently shutting down and starting up new businesses (again making investigations slightly more complicated).

This strategy was employed by one rather productive crew chief. Between 2014 and 2016, Hau operated three different bait companies, selling over

$11 million worth of worms to nineteen different coolers. Each invoice included a separate line item for HST that was never remitted to the government. In total, this amounted to $1.3 million in unpaid taxes. The crew chief could have minimized his tax owing had he documented his allowable expenses. But he did not. Nor did he keep any documents at all. The judge noted: "[Name] did not file any payroll accounts for employees with the CRA although he had employees or associates who assisted him. [Name] admitted to not keeping business records nor during the investigation were any found." Lawyers sought a conditional sentence, noting their client was "broke." His bank accounts revealed money deposited but quickly withdrawn, and he declared bankruptcy in 2016. The whereabouts of the money was a bit of a mystery, aside from "a half-hearted explanation that he gambled some of it." In the end, the man pleaded guilty and was "sentenced to an eight-month sentence and a fine in the amount of $364,000"—the amount of tax owing on just one of his three companies.[16]

Another way to fraudulently capture HST is to produce a fake invoice for allowable expenses that never existed. As the case above demonstrates, businesses in Canada with revenue over $30,000 must collect sales tax from their customers and remit it to the CRA. However, businesses can reduce the HST they have to pay by deducting the amount of HST *they* had to pay on expenses necessary for their own production. This is called the Input Tax Credits (ITC). Businesses will then submit the HST charged on their sales minus the ITC. Thus, one way a business can defraud the government is by creating fraudulent invoices of purchased supplies to claim their ITCs. One cooler put it more clearly: "Think about it. All you have to do is produce a million-dollar receipt and you get $130,000 back. That's what happens." How does this work in the bait industry? For the cooler who buys, let's say, $1 million worth of worms, they will pay an additional $130,000 in HST to the crew chief. When they export those worms to the United States, they submit a B13, which specifies the dollar value of the worms and taxes charged. This document will allow them to reclaim the HST they paid to the crew chief. The crew chief, in turn, must remit the HST to the CRA (or risk jail time and fines, like Hau). If, however, they can show they paid HST on their own operating expenses, they can reclaim some or all the HST due. Some crew chiefs and coolers who do not export are thus able to set up another nonexistent "bait company" that supposedly sold them the worms. This additional company allows the crew chief to charge the cooler HST for the worms, remit it to the CRA, then reclaim the 13 percent with another fake invoice from a nonexistent expense. Again, this form of fraud extends far beyond the worm industry.

The challenge for would-be HST-rebate fraudsters is creating seemingly credible invoices that cannot be easily confirmed as false through CRA investigations. A multimillion-dollar used-clothing import-export business that is claiming ITC, to take a real example, would not only have to document financial transactions but also demonstrate their capacity for producing and stocking the quantity of product necessary to justify the expenses. In this case, when the CRA conducted an "inspection of the business premises," it became clear that it had been "claiming refunds for purchases that were either fictitious or not incurred by the company."[17] But how would the CRA inspect a crew chief's premises? The material production and capital investments of the nightcrawlers and capital investments allow crew chiefs to mix legitimate and illegitimate expenses to easily justify invoices. Worms—the taxable commodity—are only in their possession for several hours in the morning, essentially the time it takes to collect the worms from the pickers until they are dropped off at the cooler's. They do not stock them. It is also normal for several middlemen—be they crew chiefs or other drivers—to buy and sell between each other and the coolers.

Added to the complexity is the nightly variation in the number of worms picked, given the influence of variable weather and picking conditions. There is no "normal night" of picking. A crew chief might collect half a million worms in a night, or maybe just 10,000. Adding hundreds of thousands of worms on a fake invoice would not send up any immediate red flags. Working with over a dozen coolers and other regional middlemen, therefore, makes it difficult for an auditor trying to distinguish between a real and a fake invoice, real or fake companies, all of which could feasibly be justified based on the normal functioning of the business.[18]

Coolers will not hesitate to call the CRA if they have knowledge about their competitors' potentially illegal actions, another reason to remain tight-lipped. "He'll [a competing cooler] use the information to get you down. If you leave him alone, he'll leave you alone. If you don't, he's going to give some info to the tax people—HST. It's amazing how much he knows about everybody else's business."

For all those who want to *avoid taxes* on otherwise legal activity, there are others who want to *pay taxes* on otherwise illegal activity. Money laundering is simply taking money obtained illegally or illegitimately and making it appear as though it came through legal activities. In the course of my research,

I heard accusations that the Italian mafia as well as various motorcycle gangs were able to use the nightcrawler industry as a means to launder money. It is, after all, well suited to this task. First, it is primarily a cash business, which theoretically allows easy manipulation of accounts. Millions of dollars of cash flow up and down the value chain between pickers and crew chiefs, crew chiefs and coolers, farmers and crew chiefs, as well as farmers and coolers. But many other transactions are done with checks, and it would not be difficult to mix clean and dirty money.

The process, I was told, would work as follows. Some organized-crime syndicate needs to launder $100,000. They create a bait company, with a registered HST number, and produce an invoice for the cooler. The cooler, who is likely paying cash to other crew chiefs in exchange for *actual* worms, uses this fake invoice to cover the transaction that had no documentation. The illegitimate money is cleaned, invoices are in order, and taxes are paid. Even if the total quantity of real worms and nonexistent worms don't match—that is, if invoice payments are far higher than the total number of worms housed in the cooler—there is a valid excuse: worms are living commodities that can easily die. One could justifiably claim that the worms were picked at too high a temperature, that pesticides impacted the worms, that the peat moss was poor quality, that the cooling system malfunctioned, and so on. Each excuse is potentially valid.

An extra perk of the nightcrawler commodity for laundering money—in contrast to other commodities—is that when it dies, it will quickly and literally disintegrate into nothing. The evidence (or lack thereof) vanishes. One interviewee described it as follows: "You can't see where the money has gone. And it's big numbers, so it's easy to 'flow' things. Worms go in the shit pile, or worms die. There's no physical . . ." He stopped to rephrase. "You can spend a lot of money on something that's not there." The going rate for laundering money, I was told, is 5 percent of the total amount of money needing to be cleaned. Obviously, this fee is paid in cash.

The connection between the worm industry and organized crime came up late in my fieldwork. Some coolers claimed they had never heard about money laundering or organized crime, while others said little more than, "Yeah, I heard about that." As such, I am unsure of the directionality of the relationship—that is, who contacts whom? Do coolers and crew chiefs offer their services to organized crime, or does organized crime seek out willing partners? Once connected, however, it appears the relationship can be reciprocal. The coolers and crew chiefs are useful for laundering illicitly obtained money, but the strong-arm tactics of organized crime can be useful to the bait industries in euphemistically "applying pressure" to competitors and indebted crew chiefs. One person described an example like this: "[Person A] sent the bikers over there to [Person B]" to ensure access to a certain piece of land. "But [Person A] owed [Person C] some money. Didn't pay. Then [Person C] sent the Italians that he had connections with over to [Person A]" to collect the money. Loaning money between coolers and crew chiefs does not appear to be rare, and these informal transactions need to be enforced. I asked one cooler about a company I'll call HW Bait Company. The cooler laughed. "He's not in it anymore. [Name] was going to beat him up. Owed him $10,000." The cooler then asked me, "Did you see [name] walking in a cast? They [bikers] broke his leg . . . he owed them half a million bucks." The cooler stopped to clarify: "Apparently. I don't know."

The literal and figurative underground nature of this industry allows some businesses to capture more of the surplus because of how earthworms are formally subsumed; those who avoid taxes and/or launder money retain more profit than those operating legally. Any nightcrawler capitalists paying cash, laundering money, or skirting HST payments retain surplus value that can be spent on renting more productive fields and paying pickers higher wages. Contrast this with those coolers who claim to be working according to the law, writing checks, organizing payroll, and paying HST. They are clearly aware of their own impairment. "You might make a bit more money in the worm business if you don't have to compete with those guys. It's tough to do an honest job and have to deal with idiots like that."

Many of the crew chiefs and pickers will not even accept payment in checks. "A lot of the other coolers is cash, cash, cash," said another cooler, and if they don't pay cash, they risk losing the crew chiefs and pickers to other coolers who will. Nico tried to put pickers on payroll, but they refused. "Okay," he said, "give me your ID because I have to give you the check. 'No. You have to pay me cash,' otherwise, she'll quit and go with somebody else." Other coolers complained as well: "I pay my taxes. I paid $80,000 to $90,000 in taxes a year. They [other coolers] don't pay nothing, and they got twice bigger business than me." This cooler said he would like to have an "undercover cop, one night, with me in my car, and we're gonna go from field to field, and you're gonna see the guy carry $10,000 in cash paying pickers right there." Unfortunately, the cooler told me, the "CRA says, 'we can't monitor the industry because it's always at night'" and spread all around the countryside. "They don't have the manpower to check out the addresses that I gave them." For as much as the industry has formalized throughout the decades, he told me, still, "it's gotta clean up."

· · ·

Nightcrawler capitalists are stuck with the resources available to them. Aside from the brief example of Nitro Worms, there have been no breakthroughs in taking hold of the production of nightcrawler worms nor the conditions of their production and reproduction. For more than eighty years, no technology, no purposeful application of science, nor any amount of cooperation has significantly changed the production process. The low barriers to entry and the ease of entering at various points in the production chain create fierce competition where crew chiefs and coolers undercut one another in the context of a dwindling market. The formal subsumption of labor and nature and the materiality of nightcrawler physiology and ecological function allow many aspects of the production process to evade formal institutions. The undergroundness of the industry enables some nightcrawler capitalists to outmaneuver their competitors and pocket the surplus before it is lost to the state. How are these additional surpluses spent? They are used to procure the rarest of commodities in the industry—labor power.

FIVE

The Worm-Picking Labor Process

The first time I went to pick worms, I couldn't stand. I couldn't sit. It was really painful. It's like your whole body is aching and pain. After I take a shower, I have to go to bed. I have to pick up one leg and then pull up the other leg because it hurts a lot.

SUNI, worm picker

At night, when other people are with their families, you're on the field, where its cold, dark, dirty, under the rain, the fog. Sometimes it's so cold your hands are paralyzed.

THE WORM QUEEN

IN A CHIC DOWNTOWN TORONTO CAFÉ, the seventy-year-old Vietnamese woman squatted on the ground and proceeded to—as they would say in the fitness industry—walk like a duck. Her bent knee rotated to take a step, followed by the next, repeating as her body waddled forward. She stood back up. "I can still do it, but not much." Worm picking, she said, was exhausting, "Nothing compares to the harshness of it." I asked for a comparison nonetheless. She said in the late 1970s, as border skirmishes along the North Vietnamese border with China intensified, the Vietnamese government expelled her and other ethnically Chinese people to China. There, she worked on a government-run sugar plantation and then transferred to a rubber plantation, getting paid about a dollar a day. "And that," she said, "was not as hard as picking worms." Her friend, a woman about the same age who picked worms in the 1980s, concurred. "Even in Vietnam, when you're working on the field, it's not this hard work." I asked if there was anything about the job they missed. Their answer was short. "Nope." They laughed together.

A recent job posting for "worm harvesters" describes the work environment as "Outdoors; Wet/damp; Odors; Dusty; Hot." The labor process includes: "Repetitive tasks; Physically demanding; Manual dexterity; Hand-eye coordination; Standing for extended periods; Bending, crouching, kneeling; Large workload." Worm picking is not an attractive job opportunity, and few people are willing to strain their bodies in farm fields in the middle of the night. However, changing these conditions to find more laborers would require more control over the ecological conditions of production. If one could control the soil, the temperature, the moisture, and the feed, and even select for genetics, as can be done with red wigglers or European night-crawlers, the labor process would radically change, and the pool of laborers along with it. Instead, the nightcrawler capitalists are stuck dealing with the less-than-ideal biophysical particularities that persistently inhibit productivity gains and fundamentally shape how labor is articulated in production. The uneven ability of nightcrawler capitalists to subsume the worm shapes the constitution, composition, and conditions of labor.

This chapter details the worm-picking labor process to see concretely how nature shapes who labors and how they labor. I begin this chapter with a description of the working night to detail the concrete labor process of worm picking. Worm picking remains a "traditional" or "archaic" form of labor, having changed little from the early description of late nineteenth-century worm pickers scouring city parks. I describe picking equipment and techniques as well as the physical and mental toll of picking as many worms as possible under challenging conditions. In a way, there is no special skill to worm picking; it has been done by men, women, and children of numerous nationalities and cultural backgrounds throughout the decades. And yet, while any able-bodied person can pick worms, few do. Today, the majority of pickers are Southeast Asian immigrants, primarily from Vietnam, now in their fifties and sixties. Why is this the case?

Seeing how the subsumption of nature and labor are conjoined reemphasizes the contingent nature of assemblages necessary to produce surplus value. Different bodies and different landscapes are necessary at different times to maintain an acceptable value of labor power and permit continued accumulation. How this has (recently, and temporarily) stabilized in the worm-picking industry is the result of multiple processes operating across scales: the geopolitical forces and colonial histories that spark waves of immigration, state institutions that regulate visas and minimum wage rates, household responsibilities, and, of course, the laborer's own physical capabilities and mental stamina. Particularly where capital is stuck in formal subsumption, it must thread together the less-than-ideal forces of production not of its choosing and attempt to enroll certain people at certain times in a labor regime co-constituted by the worm. Assembling labor around the worm is not easy, and it gets more difficult by the day.

As the sun began to set, the rural dirt roads were quiet, aside from a few tractors I passed. I arrived at one of Harrison's fields around 9:00 p.m. Since we last spoke, I found he had increased his worm acreage from 90 to 120 acres and increased his rental rate a hundred dollars to $1,100 per acre. He had also moved into a new home, which was surrounded by one of his new worm-picking fields. There were two vans parked alongside the gravel road next to this property. A small group of worm pickers had seemingly come to the farm early and laid out two blankets at the edge of the field; the women were sitting in a circle talking while the group of men beside them were playing cards. As I introduced myself, they directed me to one woman designated as the best English speaker. I asked when they would start picking worms, and she said they might begin at midnight. What were they doing there so early?

Harrison texted me to say he was off to bed, as "5 am tends to sneak up quickly," but I was free to linger on his land as long as I wanted. By 10:00 p.m., his house was entirely dark and quiet. There were no cars on the road, and with overcast skies, the field was completely dark—save for a few headlamps the people at the edge of the road had turned on. It looked and sounded like a campsite, with the chatter, laughing, and singing cutting through the silence of the night. For piece-rate workers, they did not appear particularly anxious to maximize their working time.

At 10:05, I heard a distant rumble, and soon a vehicle appeared, speeding along the dirt road approaching the field. It passed the worm pickers sitting on the blankets and turned onto a pathway to drive deeper into Harrison's field. Every few minutes, I heard more rumbles and saw more vehicles approach. By 11:00 p.m., some of these newly arrived pickers had donned their headlamps and begun checking Harrison's field, but were not yet picking. At 11:15 p.m., however, I checked my phone and saw the temperature had dropped to 17°C; it was safe to start picking the worms. The pickers must have checked their phones at the same time, as within minutes, dozens of people with headlights had spread across 50 acres of land, and the working night began.

The length of the working night is determined neither by capital nor by state regulation but by the behavior and physiology of the worm within its habitat. Pickers start picking when light, temperature, wind speed, and moisture align, and cease picking when the sun's rays illuminate the sky at dawn. As these conditions fluctuate both daily (temperature, precipitation, humidity) and

FIGURE 12. Worm pickers collecting worms on a farmer's field. Photo by author.

seasonally (temperature, time of sunset), each working night will have a different start and end time, which significantly constrains the nightcrawler capitalists' ability to extend the working night and increase absolute surplus value. As such, the character of the production of nightcrawlers is similar to the production processes of those late nineteenth-century men and boys hunched over and stalking nightcrawlers in city parks with a tin can and lamplight. Although there is "no essential change ... in the real production process," Marx notes, "these modifications" of the production process "can only be the gradual *consequences* of the subsumption of given, traditional labor processes under capital, which has already occurred."[1]

Worm pickers today continue to bend, hunch, squat, and crawl. They continue to pick at night and use lights to illuminate the moist ground. The push to increase productivity has transformed the type of land and organization of capital as a means to intensify production without significantly transforming the labor process itself. There have also been slight changes to the picker's technology and equipment. These "gradual consequences of subsumption"[2] create minor increases in productivity that are quickly adopted by the majority of pickers until they become generalized throughout the industry. When wages are paid per thousand worms, any minor change in productivity results in more money for the picker and capitalist alike.

One of the women sitting on the blanket provided me with the necessary equipment to pick worms. She first handed me a headlamp, but I told her I had come prepared, holding up my cheap camping LED headlight. The woman shook her head and instead gave me a dimmable headlamp attached to a lithium battery pack. LED lights are a vast improvement over the kerosene lanterns of the late nineteenth-century pickers and also much cheaper and more efficient than the heavy miner battery packs that were prevalent in the industry between the 1950s and 1990s.[3] Historically, the miner headlamps and LEDs cast a white light on the ground,[4] but this presented a problem. Shining the harsh white wavelengths on the ground hits the photoreceptors in the eyeless worms' skin, triggering them to retreat into their burrows. One picker I interviewed told me his strategy was to shine the light slightly askew of where he wanted to pick to prevent startling the targeted worms. Most pickers now use red light, which produces much longer and less disturbing wavelengths. Red lights, however, cause more strain on the pickers' eyes. When I turned on the headlamp the woman had given me, I was surprised to see it cast an orange light. The Thai picker who gave me the light suggested the orange light was a compromise: less triggering for the worms than the white lights and easier on human eyes than the red. All the Thai pickers were using orange lights, while all the Vietnamese had red lights.

I was expecting the pickers would give me two cans and some pieces of inner tube to tie the cans around my ankles, like the 1959 photo reproduced as figure 13. I had often read how one can is used to collect worms, and the other would hold sawdust that would dry my fingers after clasping a slimy worm. But these pickers had decided to go with a newer system that was catching on—a system whose apparent inventor was none other than the ultra-productive Worm Queen herself. In our interview, the Worm Queen described herself as a "small lady," and she found it painful to tie the cans to her ankles. It was also tiring to lug around several pounds of worms and sawdust attached so low on her body. Her idea was simple. She sewed some straps together to customize a harness that would go around her shoulders and clip to the worm can at her torso. For the sawdust, she created a "pouch" that would attach to her belt. No more bruises or scratched ankles, no extra weight so low to the ground. She could move "softly" with quick steps—another crucial strategy to prevent worms from retreating. Other pickers told me how this minor change, removing cans from ankles and affixing them to their torsos, redistributed weight, reduced pain around the ankles, helped with endurance, and allowed smoother, undisturbing steps. Another man I spoke with had adopted this strategy, noting he could pick more worms because the "ground will be quiet." Granted, these are not revolutionary inventions that

FIGURE 13. A worm picker in 1959. Used with permission by *Montreal Star.*

have changed the fundamental character of worm picking. Workers are still on farmer fields in the middle of the night, bent over, and pulling worms from the ground, but this slight modification has slightly increased both productivity and comfort. Indeed, the benefits were obvious to fellow worm pickers, and the technique appears to have spread between crews. Another female picker who recently retired told me how pickers watch and learn from one another: "Everyone has their own style, and one person will copy another." The Worm Queen was adamant in our interview, however: "I made it. Some people copy me." All the Thai pickers and most of the Vietnamese pickers I saw on Harrison's field were employing the Worm Queen's modified equipment, again pointing to the level of autonomy and control workers possess over a labor process characterized by formal subsumption.

The Thai woman gave me the Worm Queen–designed harness to carry the tin can—black polypropylene strapping handsewn into two connected triangles to fit over the shoulders. Two carabiners attached to the front of the harness snapped onto metal eye hooks attached to the tin can. She also provided a pouch, about the size of a photographer's lens holder, that was filled with fine sawdust. Last, I was given a small backpack filled with handsewn mesh bags. Once I filled my can with worms, I was to dump them into a bag and leave them on the ground for the rest of the night, and gather all my bags in the morning. There was a small slit at the bottom of the backpack that allowed me to pull out the mesh bags one after another like tissues from a Kleenex box. There were a lot of bags in the backpack, as the woman was probably capable of going through three or four per hour. I would only need one.

"Okay," the woman said, "go pick." I was expecting a few more instructions. "Where?" I asked. She looked at me like she didn't understand my question. "Go pick," she repeated simply. I shone the light on the ground and immediately saw why she seemed confused. The worms were everywhere, under my feet the whole time, quite literally there for the picking. Some worms were merely peeking out of their burrows; others were stretched out across the ground. I observed some worms that appeared to be engaged in some sort of courting ritual, bobbing their heads around one another before sliding their slimy bodies against one another to begin copulating.

I took my first step and bent down to pick a worm, but the strength of the orange headlamp was magnified, and most of the worms disappeared in an instant. I had been warned that the worms were "very fast" that night but was still surprised—and frustrated—at the speed of their disappearance. The only worms remaining were the copulators; recall how Darwin, after one of his dew worm experiments, noted how "light would not disrupt their sexual passion." Grabbing them, however, certainly disrupted their sexual passion. They quickly detached from one another and tried to retract their bodies back into their respective burrows, but it was too late for them. Throughout the night, I was delighted to find the copulating worms—a two-for-one opportunity I would not let escape.

In some ways, the physicality of the labor process is similar to something like strawberry picking, where crouched speed and rapid hand movements are paramount to increase productivity. But there are some "lively" differences. A strawberry is on a stem and doesn't try to retreat into a hole when touched. And though strawberries are considered a tender fruit, I would suggest they are not as tender as a fleshy, stretchy worm. I was conscious not to "pinch" the worms—squeezing them a little too tight—as I had witnessed

FIGURE 14. Photo of author with front-mounted can. Photo courtesy of author.

FIGURE 15. Bag of approximately 300 worms. Photo by author.

victims' limp gray bodies at the coolers' warehouses. But it is rather difficult *not* to pinch a slimy, boneless creature with such a remarkable grip on the soil. When I could get a decent grip, the worm would come out with the sound (and satisfaction) of uprooting a weed. Other worms could hold to the soil so tightly that I would tear their bodies in half. I probably severed (although did not necessarily kill) about a dozen worms throughout the night.

I began the night squatting and tried to copy the movements of the pickers I saw around me. "Walk light and pick fast" was the summation of the Worm Queen's strategy. But this required an agility that I did not have. If I stepped too clumsily, the worms disappeared. If my toes dragged a few inches, they would be gone. I looked at another picker who was much more successful. He was not as fast as I would have expected for someone on pace to gather 20,000 worms, but he was moving steadily and smoothly. His hands operated in asynchronous motion, each pulling one, two, or three worms at a time. One after the other, the picker's hands would drop worms into the can, plunge into the pouch of sawdust and attack the ground again, all while methodically taking smooth and soft duck-like steps forward to repeat the process again. Several of the interviewees described similar picking strategies. The Worm Queen told me she would grab at least five worms in one hand before dumping them in her can. "Some people pick worms like a vacuum," said a picker whom I will call Nguyen. "Or maybe they have a vacuum in their hand! Picking several worms, two hands at the same time. . . . Not one by one." Another picker told me that once the worm is in your fingers, you have one chance to pull in the same direction as it is angled in its burrow: "Once you tug, you have to get it." Picking too slow would also allow the worm a secure hold in the burrow and the ensuing tug of war would rip the worm in two.

Another Vietnamese man in his fifties told me there wasn't any particular technique in picking worms. Rather, success "depends on your back." When his back would get sore, he would return to squatting until his knees would get sore, at which point he would revert again to bending. Another picker told me that was his strategy as well: "Yeah, that's the worst. . . . When it's sore here [pointing to his back], I squat." I employed this strategy throughout the night, sometimes squatting, sometimes bending, and other times, out of exhaustion, simply crawling on my hands and knees (which was a terrible strategy for not disturbing the earth). As the night went on my back and knees began to ache, as I had anticipated. Nguyen told me, "When I squat and it's sore, I bend. And finally, sit down, smoke, and take a rest for a couple minutes." It was approximately an hour and a half after picking started that I got the first whiff of cigarette smoke.

Mai, in her fifties, briefly picked worms in 2018. She remembered her first night picking clearly. "I heard from my friend, 'Oh it's an easy job, you can make money to buy a car.' And I thought, 'Oh, they can do it, so why can't I?' I thought it was simple, but it turned out . . . not so." She was not appropriately dressed, wearing a leather jacket and jeans. She remembers her headlamp fell off her head and all the wires got tangled. In the darkness, trying to reassemble electrical connections, she began to cry. She picked 1,000 worms that first night. Her second night, she picked 2,000. But that was it. She told the driver she would not go again: "They asked why. I said my whole body ached everywhere. They said, 'You have to try to get over yourself to make it a habit and you will get over those aches and pains.' I said, 'No.' I quit after those two nights."

Another woman, now in her sixties, told me a similar story. Her friends told her about all the money she could make, but after two nights, she was done. Recalling those two nights, she said, "There is nothing good about it. It's only for people without a job." The first few nights were always the worst. "The first time I went to pick worms, I couldn't stand. I couldn't sit, I couldn't stand. It was really painful," said Suni. She remembered having to crawl up the stairs and literally picking up each leg to get into bed. "My whole body just burned. It hurt."

Coolers and pickers both note the high turnover rate for pickers, suggesting it takes at least two months to become productive and profitable while weeding out the less productive pickers whose physical exertions do not reap high wages. Several of the pickers noted how their bodies got used to the rigors of picking after several weeks, but year after year, they developed chronic pain, specifically in their knees and back. As one Ontario Workplace Safety and Insurance Board ruling notes, "The job of worm picking, as described, is a job that may well cause a back injury, particularly to someone who performed this job over a number of years, as the worker has done. . . . Undoubtedly the Board has come to realize that the worker's compensable injury has had much to do with the condition of her back."[5]

One woman said she could no longer squat without physical pain. Another man said he stopped at age fifty when his back could no longer handle nightly picking. Among the pickers I spoke with, back and knee pain were consistently cited as the worst part of the job and often the primary reason for quitting when other opportunities arose. "Even me, coming from Vietnam," said one woman, "it made me cry certain nights."

When the temperature dropped to 11°C, I started to get chilled. I had also been awake for about thirty hours at that point and fatigue set in. The worm pickers who had loaned me the equipment guessed I would get one can (about 300 worms) for the night, and I decided I would not disappoint. At 4:00 a.m., I reckoned my can was full. I dumped it into a mesh bag, went to my car, and took a twenty-minute nap.

As the sun started to rise at 5:00 a.m., the birds started singing, and the picking slowed. Several worm trucks—which were not used to transport these pickers—started rumbling down the rural road and drove onto the field. The drivers opened the back of the trucks and started to unload hundreds of white flats. A few pickers were still squatting, grabbing the worms in front of them. Other pickers were standing up to survey the remaining worms, only bending down to grab one from time to time. Others were criss-crossing the field to pick up their mesh bags. As I was walking around, trying to remember where I had left my mesh bag in this 50-acre field, I passed the

FIGURE 16. Worm pickers loading worms onto truck. Photo by author.

Thai picker who had spoken with me earlier. She was surprised to see me. "Oh, you're still here? I thought you went home." I said I did not, but had only managed to pick one can. She said, "Yeah, it's not a good night. The worms are very fast." I asked how many bags she picked. "Only seventeen."

Crew chiefs brought four-wheelers to crisscross the field, picking up mesh bags that they knew belonged to their pickers. As they returned to the truck, each picker found their mesh bags of worms and dumped them into the driver's flats. Drivers recorded the number of flats each picker had filled and loaded them onto the back of the truck. The pickers I worked with left in their own cars. The working night was done.

PIECE RATES

If I had been paid for my 300 freshly picked nightcrawlers, I would have made about $7—exactly $0.023 per worm. Had I been paid the hourly Ontario minimum wage for my time picking worms—about four hours, excluding my nap in the car—I would have been paid $57, or $0.19 per worm, and, no doubt, would have been immediately fired. Piece rates are critical for nightcrawler capitalists to attract productive laborers and maintain profit margins. Marx notes how piece rates "furnish to the capitalist an exact measure for the intensity of labor,"[6] and, unsurprisingly, contemporary economists, "in particular, have a predilection for piece rates,"[7] as they provide a specific quantitative measure value while often increasing productivity and reducing supervisory costs.[8] For the nightcrawler capitalists, their rate of profit is clear, embodied in each worm whether it was picked productively by an immigrant or unproductively by an academic. They do not pay for the time pickers take in transit or their time waiting for the temperature to drop. Nor do they pay for each sip of water or puff of a cigarette. Piece-rate workers internalize workplace discipline and are incentivized to increase intensity as well as the length of the working day to increase their wages. This form of "self-exploitation" often takes a greater toll on worker bodies while undercutting collective forms of worker solidarity.[9] At the same time, in contexts where labor is scarce and workers can change employers, piece rates can enable diverse forms of worker agency. Particularly, fit and productive workers can command higher rates while not tied to a wage contract, which might seem "galling" for productive workers.[10] Gidwani also stresses how the prevalence of piece work in agriculture includes "place-specific cultural valences that become attached to work" and how these shape worker identities and the terms of employment.[11]

Only implicit in the agricultural piece-rate literature, however, is how control over ecological conditions of production shapes the formation, dissolution, or maintenance of piece rates. Not every production process can easily be broken down and quantified into abstract "pieces" or a narrow set of repetitive routines,[12] nor are they all static over time; transforming the production process through certain technological advancements will disrupt the previous forms of compensation and lead to new payment schemes.[13] In the conceptual terms of this book, the transition to real subsumption often involves adapting or jettisoning previous payment schemes.

In the worm-picking industry, the (lack of) control over the ecology of production necessitates piece work for both labor and capital. Both pickers and coolers laughed when I asked if pickers would work for an hourly wage. Pickers said it would unfairly punish the productive pickers and reward the unproductive. Piece rates also allow the pickers to work only when they deem it will be profitable for them. They do not show up at predetermined times but rather carefully monitor weather conditions on their phones and judge for themselves whether traveling to the farmers' fields will be worth the time and effort given prevailing piece rates. Once in the field, conditions may change in the middle of the night and pickers may decide to drive to another field or return home early to avoid morning traffic, get extra rest, or prepare lunches for their school-age children, or often because they have made enough money for the night. When the price of worms was high in the 1980s, for example, a woman remembered picking $50 worth of worms in a couple of hours. That was all the money she needed that night, and she promptly went back to the truck and got a good night's rest. When I talked to the Worm Queen early in the picking season, she noted the piece rates were only $18 per thousand worms. Combined with the low temperatures and moderate windspeed, she suggested it would not be worth picking for at least a few more weeks.

Pickers also mentioned this as a benefit, as they see paying equal wages in a physically punishing labor process to be unfair. Piece-rate wages created a meritocracy in the field that no one would exchange for paltry hourly wages. For the Worm Queen, "Your income is entirely dependent on your ability.... Your earning depends on you." It's quite simple, Mai told me: "The more you try, the more money you can earn." Two pickers told how they would often imagine that the worms were coins lying about the ground. "Back then," a retired picker told me, "People would say you had one cent per worm. So, I just think I'm picking one cent, one cent, money, money, money." A similar sentiment was shared by Suni, who noted when the worms are out in full force, the back pain and knee aches seem to magically disappear: "You forget. You just work. It's like money. Money, money, money. You just don't think about anything, just money, money, money." Most pickers viewed this as "fair," with one picker laughing at the idea of being compensated by the hour.[14] She got her share, but so did crew chiefs, coolers, regional wholesalers, and retailers—each of whom takes a slice of the surplus labor embodied in the worms.

WHAT MAKES A GOOD WORM PICKER?

John Reynolds, the earthworm taxonomy expert, first heard about the worm-picking industry as a teenage counselor-in-training at an Ontario summer camp in the 1950s. He remembered there was a senior camp counselor who decided to quit and make more money worm picking in the summer, which in turn promoted Reynolds to full camp counselor. In the early days of the worm-picking industry, worm pickers were often born in Canada, including people who worked middle-class jobs throughout the day. One older cooler remembers that some of his first employees were low-paid Canadian naval mariners who never saw any wartime action. They would pick four or five nights a week, grab their paychecks, and disappear into the local bars: "They were literally drunken sailors." The piece-rate wage could be so high in the 1960s that he began buying worms from Air Canada employees who would make $28 at their daytime union job, and then an additional $70 picking worms at night. He also drew the attention of one particular policeman who showed up at his warehouse one early morning, not because of any investigation or warrant but because he had a trunk full of worms to sell. I also spoke with a white Canadian-born man, now in his fifties, who picked worms in the 1980s on golf courses. He was never committed to the job but would just show up whenever he needed to make $10 for his teenage escapades. However, from the late 1960s onward, this became extremely rare; the primary commonality between pickers is their status as immigrants.

Today, pickers, coolers, and crew chiefs all note that most pickers continue to be of Vietnamese origin, estimating that they make up between 60 and 90 percent of the workforce. It was implied to me, sometimes hesitantly, sometimes with confidence, that race and culture played a role in making a productive picker—that somehow there was some kind of "natural" advantage or cultural predisposition that made Vietnamese people into productive worm pickers. Is this true? Are Vietnamese immigrants naturally good worm pickers? Or does the industry naturalize Vietnamese immigrants at worm picking?

Capital accumulation depends on inequality to accumulate—both the inequality between owners and workers and that between workers themselves. Certain bodies are socially constructed to be undervalued and/or unvalued to ensure capitalists maintain low wages and, subsequently, higher profits.[15] This results in labor markets that are structured around differences in class, race, ethnicity, and gender.[16] Particularly with formally subsumed labor, the production of racial and gendered hierarchies naturalizes and justifies that certain people perform certain work, whether it is manual labor, care work, reproductive labor, or piece-rate jobs. The process of racialization in particular "enshrines the inequalities that capitalism requires."[17] This is especially true for immigrants and migrants whose skin color intersects with language abilities, ethnicities, and cultural attributes that signify what types of work they are expected to perform.[18] Certain people are devalued in order to produce value, and this is evident in worm picking. The demography of worm pickers is neither natural nor coincidental but a result of systemic factors that uphold economic and racial inequalities.

Racial hierarchies are rampant in worm picking. Cooler and pickers alike were usually not afraid to assert how Vietnamese were best because of their physical characteristics or cultural background. But the more we unpack worm-picker backgrounds, the more these naturalistic and Panglossian assumptions about what makes a good worm picker began to break down. The Worm Queen suggested Vietnamese were "slimmer" and also "built for endurance," and "closer to the ground," making it easier for them to squat and move around. But others noted that Vietnamese *and* Greek pickers were the fastest. Another picker remembers a Hungarian woman being the most efficient back in the 1980s, while in a 1972 newspaper article, Bob Druin notes how one picker harvested 22,500 worms in a single night—a phenomenal yield considering the time. That picker was a Greek man. Refugees from Kosovo were also deemed productive pickers, and now most coolers and some pickers suggest Thai pickers are the fastest.

If it wasn't size, perhaps it was more about culture. Coolers and pickers suggested the Vietnamese "have a history working in fields," particularly planting and harvesting rice—perhaps one of the concrete labor processes most analogous to worm picking. Two other female pickers I spoke with shared this sentiment, with the added observation that the best pickers tended to come from North Vietnam, which had a higher agrarian population. But this observation didn't track evenly across the industry either. The Worm Queen was a student from South Vietnam. Nguyen was a machinist in Ho Chi Minh City. Suni did work on a family farm but not in any kind of labor-intensive manner. The same holds for the earlier Greek pickers, as well as contemporary Thai pickers I spoke with. One had worked at a vegetable processing facility, another drove moto-taxis, and another was a truck driver.

DCA Baits also scoffed at these naturalistic narratives and suggested it came down to individual "drive." Instead, the characteristic of a good picker had less to do with geography, experience, or physical characteristics and more to do with the individualized concept of "work ethic" combined with a "fit" body (as compared to a "small" body). A Vietnamese cooler suggested this had something to do with Vietnamese notions of hard work. "They [the Vietnamese pickers] see the value in the worm . . . and are willing to work for that value." DCA Baits, however, didn't completely agree. He suggested it was less about some cultural predisposition to hard work and more because of their history as refugees, having gone through unimaginable difficulties that provide the necessary "drive" to persevere under the grueling picking conditions. He summed it up clearly: "If you are coming from a civil war, you can do a lot of things."

But again, being a refugee fleeing civil turmoil is neither a necessary nor a sufficient condition of picking. One woman I talked to, now in her sixties, came to Canada as a refugee in the 1980s and started to pick worms because she wanted the income. After two nights, she quit and never returned. In contrast, the Worm Queen was not a refugee, nor are other pickers who travel to Canada on visitor visas. Nor, for that matter, have more recent refugees engaged much in the worm-picking industry. Over the past decade, Parga Wholesale Baits has specifically recruited Colombian, Ethiopian, Pakistani, and Syrian refugees through advertisements in community newsletters and online ads. Few, he said, ever lasted more than one night, let alone a whole season.

There also was some speculation that worm pickers tended to be women and that women are physically and mentally better suited for picking than men. There is much academic literature that critiques such essentialist characteristics of gender in the labor process and, in turn, how the capitalist production process defines gender roles in the production process rather than the reverse.[19] The conception of a low-skilled female worker having a natural predilection for piece-rate work, possessing "nimble fingers" and passive obedience to employers, is a longstanding trope, particularly in piece-rate industries.[20] Women, it is often thought, are predisposed to meticulous, repetitive tasks because of such "nimble fingers."[21] The concept of naturalness, however, seems to miraculously fade when the wages and conceptions of the job rise in stature, at which point employees change to men (for a classic example, see "data entry" or "coding").[22]

With such conceptions in hand, however, we might assume women are more naturally suited for worm picking. Do they indeed make up the majority of pickers? As should be clear by now, a definitive and accurate demographic survey of worm pickers would be extraordinarily difficult. While some coolers and pickers suggested women are the majority (likely based on their individual observations regarding their specific crews), others suggested it was roughly an even mix. If female pickers do outnumber male pickers, the margin, based on interviewee observations, would be slim.

But are they better pickers than men? Again, I am at a loss for reliable productivity data. Interestingly, however, when coolers do have the choice to employ specific genders, nationalities, and ages, as they can do through the "agricultural stream" of the Temporary Foreign Worker Program (see next chapter), they tend to choose a mix of men and women or select only men. The perceived superior ability of women appears less to do with their gender and more to do with the structural conditions that limit opportunities for marginalized populations. As Mary Beth Mills argues, women are not inherently better at repetitive, meticulous piece-rate tasks requiring endurance in arduous labor conditions; instead, their predominance in such employment is a reflection of their social position, where particular "layers of vulnerability"—be they social roles, migrant status, ethnicity, class background, language ability, or domestic care responsibilities—disproportionately affect women more than men.[23] Nighttime, atypical, or irregular hours, for example, are recognized as "gender-coded factors"—variables in working conditions that affect men and women differently based on their social roles.[24] Worm picking requires a person to be away from their home from the late evening to the early or mid-morning. One picker noted that in the late 1980s, she was often away for sixteen hours of the day—being picked up by the crew chief at 5:30 p.m. and dropped off at 10:30 a.m. the next day. Pickers with children described the challenges of nighttime work, having to rely on a myriad of formal and informal childcare arrangements—a common challenge associated with atypical hours in various geographical contexts.[25] One woman remembers paying for overnight childcare at $10 per child. Others would rely on a spouse or family members—often their older parents—to provide childcare, prepare meals, clean, and get kids off to school. After-work hours that were spent at home were hardly conducive to quality family time, as kids would often be away at school. Suni decided to quit worm picking when she realized her child could no longer communicate in Vietnamese because she had so few hours with her kids at home. She decided instead to take a minimum-wage job that allowed her to align her hours with her children's schooling. She is now working more hours for equivalent pay to worm picking, but is able to spend more time with her two children. The Worm Queen also remembered how hard it was to stay awake when her preschool-aged kids wanted to play. Now, her kids will soon be teenagers, and she is deciding whether to quit worm picking so she can better monitor any nighttime mischief they might be getting into.

So what are the characteristics of contemporary worm pickers? The physicality of the job requires able bodies possessing a level of fitness and endurance that can allow them to withstand hours of bending and squatting. All pickers (to my knowledge) are first-generation immigrants lacking a strong command of English, with few other employment opportunities. They are both male and female, though the nighttime hours put a greater strain on female pickers responsible for home care; indeed, raising children appears to be a significant impediment to finding available worm pickers. And finally, pickers must possess a strong "work ethic"—however nebulously defined. I will not speculate on the source of such a work ethic, but only note that the ability of someone to put their "head down and git 'er done," as one cooler said, does not manifest only within particular nationalities, cultures, or employment backgrounds. But if we eschew the naturalistic explanation suggesting that Vietnamese people are somehow predisposed to worm picking, and instead see their prevalence as based on devaluing and marginalization, we are still left with the question: why are most pickers today specifically Vietnamese?

AGGREGATE LABOR POOLS

The worm-picking labor regime limits the number of potential employees to those who match the characteristics described above. The problem for nightcrawler capitalists is that so few individuals match these criteria. What is necessary, from the standpoint of capital, is to increase the aggregate size of a marginalized "low-skilled" labor pool enough to increase the fraction of people who can fulfill labor needs. Worm pickers have been Canadian-born, Italian, Hungarian, Korean, Portuguese, Greek, Vietnamese, Laotian, Cambodian, Thai, and more. They have been young and old, female and male, tall and short. They have come to Canada as immigrants, as refugees, on student and visitor visas, or through family sponsorship streams. The one commonality between pickers that does remain constant from the 1960s onward, however, is limited English skills and employment opportunities. For the southwestern Ontario bait industry, the political turmoil following the Vietnam War vastly increased an immigrant labor pool that shared these two characteristics.

The end of the Vietnam War led to a massive refugee crisis, as upward of three million people fled Vietnam over the following decades. News articles and photos showing Vietnamese refugees packed onto boats headed to Hong Kong led to a public outcry in Canada. In response, the Canadian government initiated a refugee "matching" program in 1979 that would see it accept one refugee for each refugee privately sponsored by churches, community groups, or groups of individuals. In 1979, Canada granted asylum to 5,000 Vietnamese refugees. By 1985, this number had risen to more than 110,000 refugees, with over 23,000 settling in the Toronto region.

Only 12 percent of Vietnamese refugees who came to Canada in the late 1970s and early 1980s could "minimally" communicate in English or French. Nor did their employment background—often as small farmers, military personnel, or handicraft producers—easily fit into Canada's occupational categories or economic needs.[26] Compounding these challenges, refugees in the early 1980s entered Canada in the middle of an economic recession with high inflation and interest rates and double-digit unemployment. Indeed, "investigations into the initial experiences of Indochina/Vietnamese refugees arriving in and after the late 1970s suggest significant difficulties integrating into the Canadian labor market."[27]

Some of the pickers I spoke with were able to find work in some formal economic settings, often in Vietnamese restaurants where language and integration were not significant barriers. Others worked in the struggling manufacturing sector or agriculture. While they described such jobs as physically "easier" than worm picking, they were not necessarily "easy" full stop. Factory and agricultural work required significant interaction with supervisors, often involving a diversity of tasks that had to be explained in English. Research has also documented the overt and subtle forms of racism that the recently resettled Southeast Asian refugees faced daily, working alongside their Canadian-born coworkers.[28] In contrast, worm picking offered a more solitary endeavor outside the formal sector, with few words ever needing to be spoken. The pay was also high, $20 per thousand worms, especially compared to the minimum wage at the time, which was much lower than equivalent rates today.[29]

Up until the early 1990s, Vietnam continued to be the leading source of refugees and immigrants into Canada, with the 1996 census recording 41,000 people of Vietnamese origin residing in Toronto, the largest Vietnamese population in any Canadian city.[30] Over the years, worm picking became a recognized and socially accepted opportunity for Vietnamese refugees, and familial and friend-based networks of recruitment developed.[31] One woman I talked to noted she had initially settled in Edmonton but was quickly encouraged by friends to move to Toronto and make more money picking worms. All the pickers I spoke with started picking because of a friend or family member. A social knowledge of the industry developed to coordinate pickup/drop-off locations, provide picking tips, and recommend which cooler or crew chief to work for and which to avoid, as well as to offer words of encouragement to keep picking despite the aches and pains in their bodies. One man I spoke with said it was common knowledge in the Hong Kong refugee camp he was in that one could make a lot of money in Toronto picking worms. For the bait industry, the influx of immigrants provided a new and large labor pool in which to find individuals possessing the characteristics conducive to productive worm picking.

The problem for the industry now is that the 1980s was a long time ago. After the peak of Southeast Asian refugees entering Canada in the late '80s and early '90s, the rate significantly declined.[32] The only comparable wave of refugees occurred between 2015 and 2017 when the federal government resettled over 40,000 Syrians across Canada. But even these total numbers pale in comparison to the 1980s, and the available demographic information suggests important differences between the Syrian and Southeast Asian refugees. Over 45 percent of the Syrian refugees could speak English or French, and only 16 percent settled in the Toronto area.[33]

Parga Baits specifically recruited Syrian refugees and did find a few people willing to give worm picking a try. None, he said, lasted more than a few nights, suggesting that government support and increases in the minimum wage deincentivized the recently resettled refugees from engaging in worm picking. Indeed, government supports are much different from those of the early 1980s. Monthly stipends for Southeast Asian refugees in the 1980s ranged from a mere $33 to $117 (around $90 to $309 adjusted for inflation), depending on familial circumstances.[34] Recent Syrian refugees received a one-time "start-up" payment of around $2,000 to set up their house, as well as monthly assistance on par with Canadian welfare assistance. In Ontario, for example, the estimated monthly assistance to an individual refugee is $768.[35] Privately sponsored refugees, in contrast, get no dedicated funds from the government but rely on funds and support from the sponsoring group. To support a refugee family of four for twelve months, for example, sponsors would need to provide at least $28,700 of income support.[36] Whether such support is indeed adequate for refugees is another question. The point is that institutional and individual financial and in-kind support is much more established and developed than in earlier decades.

Socioeconomic conditions are also different compared with the 1980s. From 2015 to 2019 (that is, before COVID-19), for instance, the Ontario unemployment rate ranged between 5 and 7 percent. Inflation was minimal, interest rates went to record lows, and the minimum wage was much higher than in the 1980s even when adjusted for inflation.[37] Conditions within the bait industry were not very attractive either. For nearly forty years, piece-rate wages tended to fluctuate between $20 and $25 per 1,000 worms (excluding temporary increases during droughts). While it is possible to pick more worms per night on the dairy farms now (compared to the pastures and golf courses of the 1980s), this appears to have been little comfort to recently arrived refugees with no social support or industry knowledge to understand which nights to go out, which crew chief to work for, how to increase their speed, or when their backs and knees would stop aching.

In sum, the massive influx of Southeast Asian refugees, the majority of whom were Vietnamese, drastically increased the labor pool of individuals who met the criteria necessary to pick worms under the demanding labor conditions shaped by *L. terrestris*'s physiology and behavior. The social and economic conditions both in the worm industry and in Ontario more broadly created structural and racial constraints that incentivized a greater fraction of these refugees to engage in worm picking. These conditions have not been repeated in the decades since, which magnifies labor shortages today. For coolers, time is slowly running out to find new pickers. Many contemporary pickers are the *same* pickers who were working in the mid- and late 1980s, with most worm pickers now in their fifties and sixties. The Worm Queen noted she is usually the youngest picker on the field, at thirty-eight years old.

• • •

Capital's inability to control nightcrawler ecology and biology necessitates outdoor, nighttime hours, under cool and damp weather conditions, where people are paid piece-rate wages to bend, kneel, and squat while picking as many slimy, boneless, fragile creatures as possible before the sun comes up. In turn, these harsh conditions structure the characteristics of the labor force. Constrained by the recalcitrance of the lively commodity, nightcrawler capitalists must find ways to cheaply assemble productive working bodies into a labor process that most would find unappealing. Who these people are, where they come from, why they pick worms, and what makes them productive pickers are all contingent on particular criteria shaped significantly by the worms and their socio-ecological surroundings, set in particular historical and economic contexts.

The characteristics of the worm-picking labor process and control over pickers, we can see, are conjoined with the control over nature, each influencing the other. The formal subsumption of nature sets the conditions of labor, but there is nothing natural about who does the laboring. In the nightcrawler industry, economic and political conjunctures, immigration policies, and historical events have enrolled different bodies at different times. The demography of worm pickers largely mimics the flow of immigrants to southwestern Ontario from the 1960s to the 1990s, as worm picking offered relatively well-paying jobs for immigrants with limited English language skills or other job opportunities. In particular, the refugee crisis following the Vietnam War and changes to Canada's immigration policy brought tens of thousands of Southeast Asian refugees to the Toronto region in the 1980s, coinciding with an expanding worm industry in need of labor. The total number of resettled Vietnamese refugees in the region meant that capital could find the necessary laborers, ready to work under difficult worm-picking conditions, from a fraction of the total labor pool.

Such conditions have not aligned in subsequent years. There has not been another immigrant group willing to toil in the nighttime soil to replace the aging Vietnamese workforce. The persistent need provides worm pickers with a surprising amount of control over the unique labor process. When I asked worm pickers specifically about what, if anything, they enjoyed about such a terrible job, their common refrain was "freedom."

Huh?

SIX

Unexpected "Freedoms"

The weakest link being the availability of the workforce. You can control the other stuff. We can get more farms, we can spray more water, we can do whatever we need. We can change packaging.... But pickers are pickers. Labor force is a big expense.... We got no control over it.

APOLLO BAITS

If you like picking worms, it's a soft job. It's freedom, no boss, nobody controlling how much you have to pick.

MICHAEL LUNN, worm picker in the *Star Tribune*, 1988

NGUYEN CAME TO CANADA IN the mid-1980s as a refugee. He was well aware of the worm-picking industry and knew other refugees who were picking worms for decent money. But Nguyen was a trained machinist. He found a job as a welder-fitter with a good wage and steady hours. In 1998, however, he was laid off, and so he started picking worms with his wife. The cash pay was nice and comparable to his previous work, but worm picking came with another advantage. "When I [started to] pick worms, I said, 'freedom!'" He compared it to his previous job at the factory. "Every day, you have to get up early and work eight to four or seven to four," he said. "To me? Nah." With worm picking, "I don't have to get up early, and . . . in the summer, maybe I pick once or twice a week. . . . I go picking around 100 days [per year]. It's freedom." I asked what he does in the other 265 days. He smiled. "Gamble. . . . I buy and sell stocks with the money I make from picking and savings. . . . Today, win a few. Tomorrow, I lose a few. Up, down. When you study about the stocks, it's lots of fun."

Nguyen was sixty-one when I spoke with him, but, aside from the flashes of gray stubble on his face, he looked like he could have been in his early forties. He slouched back in the restaurant chair, in what I interpreted as a carefree and relaxed manner—a perception shared by the translator who was with us. For the past two decades, he and his wife have only made money by picking worms and "gambling" in the stock market. Combined, he estimated, they make over $45,000 a year (in cash), working about 100 nights per year. He knows this amount doesn't go as far as it once did, but he bought his house in the early 1990s and doesn't have a need for more money. I suggested he could sell his Toronto home, move somewhere cheaper, and retire. "No . . . I'm comfortable with my life. I don't think I'll ever sell my house." He also has no plans to officially retire as he has enjoyed the "freedom" of not having to work for someone for the past two decades.

I begin with Nguyen's account because it disrupts some common assumptions about "freedom" in the workplace. Worm pickers are mostly non-English-speaking immigrants, laboring with bent backs under dreary conditions at irregular nighttime hours, producing profit for their employers in a largely unregulated and underground industry. Such conditions do not connote being "completely free," especially for a marginalized and racialized workforce. How did these expressions of freedom emerge for Nguyen? Are these widespread among worm pickers? If not, what are the conditions that enable and differentiate such conceptions of freedom? How can we analytically resolve the apparent contradiction between expressions of freedom and a punishing piece-rate labor regime in an unregulated industry?

In a way, I would have preferred if Nguyen had not explicitly used the word "freedom," a notoriously amorphous term with different and often contradictory meanings depending on the context in which it is deployed. But he was not the only one who used that word. Nearly all of the pickers I spoke with said "freedom" or "autonomy" was the primary advantage of worm picking. Some, like Nguyen and the Worm Queen, used the word immediately and explicitly.

I found their descriptions of "freedom" most analogous to Anna Tsing's account of mushroom pickers, which is curious considering the labor conditions differ so drastically. In Tsing's account of matsutake mushroom picking, freedom comes from the escape to the forest, the idiosyncratic skill of the hunt, and a greater degree of control over one's labor. Mushroom hunting, as one hunter says, is not work: "Work means obeying your boss, doing what he tells you. Mushroom picking is 'searching.' It is looking for your fortune, not doing your job."[1]

Mushroom hunters—both white Americans and Southeast Asian refugees—find this expression of freedom in the interstices of industrial capitalism and state regulation. For some, it is even akin to a vacation from their minimum-wage work at a Walmart. Indeed, Tsing describes the mushroom pickers as forging alternate life pathways in the ecological devastation of industrial capitalism. Such precarity is both a form of survival in a neoliberal capitalist society and simultaneously a means to assert autonomy and control over one's labor.

But there is a significant difference between mushroom pickers and worm pickers: Worm-picker expressions of freedom are not derived from the concrete act of laboring. There is no escape into the woods, no thrill of the hunt, and no specialized knowledge required to pick the worms. Worm-pickers pick worms as fast as they possibly can to increase their piece-rate wages. It is monotonous, repetitive, dull, and physically demanding—quite the opposite of descriptions of mushroom picking in the forests of Oregon.

Part of the reason I hesitate to use the word "freedom" is because the racialized demography of immigrant pickers, combined with the extremely arduous and irregular labor conditions existing in an underground economy, are more representative of what scholars refer to as "unfreedom." Canadian-born citizens, after all, do not pick worms, even if they could make hundreds of dollars a night. The only reason people pick worms, we might assume, is because there are no other jobs available, and therefore, worm pickers must acquiesce to the drudging, exploitative, and off-the-books working conditions. They are free in the Marxist sense—free to sell their labor or risk starvation. But this is hardly the expression of freedom meant by Nguyen and the Worm Queen. Something else is going on.

In this chapter, I attempt to make sense of this contradiction by unpacking worm-picker expressions of freedom in the context of these unfree/free labor discussions. To resolve the apparent contradiction, I suggest that we must address how the ecological materiality of production shapes conditions and control over the labor process and the workers' experience of it. The industry has largely been stuck in the formal subsumption of nature, with the worms themselves setting the conditions of labor—the hours, the location, the picking nights, the remuneration—that influence control over the labor process. The workers retain significant control over the labor process and, as we saw in the previous chapter, are in high demand.

This is extremely frustrating for the nightcrawler capitalists. Labor is a commodity, but it is not a commodity like the compressors they use to cool their warehouses. They may need that critical commodity called *labor*, but they end up with *people* unafraid to express the small agencies worm picking affords them. They would like nothing better than to replace the labor commodity with picking machines or to develop a new nightcrawler production process. But they cannot. Instead, coolers and crew chiefs are searching for ways to revert *people* back into *labor*. The second half of this chapter describes one potential solution for industries seeking to reassert control over the laborer: Canada's Temporary Foreign Worker Program (TFWP), a guest-worker program that enables employers to hire migrants for a short period of time who are legally required to work for them, and only them. How does this program influence pickers' expressions of freedom, and does it offer a solution to the nightcrawler capitalists' labor problem?

FREEDOM AND SUBSUMPTION

Marx notes how capitalist social relations established through primitive accumulation created a "free" worker in a "double-sense": free from traditional feudal bonds or guilds to sell their labor power to any buyer, and "free" from the means of production and therefore needing to earn a wage to survive. Banaji, following Marx, calls freedom under capitalism a "legal fiction"; the choice to work for a wage or starve renders the concept of "freedom" incoherent.[2] In contrast, more liberal approaches suggest workers are "free" when they voluntarily enter contracts based on utilitarian choices available to them. Unfreedom (whether compensated by a wage or not) exists when people lack this voluntary choice and are coerced into forms of servitude.

Others see this binary of freedom/unfreedom as problematic and suggest "freedom" in the workplace exists on a continuum where labor regimes can be characterized as being more or less free.[3] Indeed all labor forms—be they paid or unpaid—"can be subject to scrutiny in terms of the extent and types of (un)freedom that characterizes them and how it structures the negotiation over conditions of work."[4] Other scholars have pushed beyond this "continuum approach" to see how "multifaceted modalities" or "axes"[5] or "varieties of unfreedoms"[6] complicate the concept of a linear spectrum that runs between "freedom" and "unfreedom." Such axes include spatial constraints, debt as forms of indirect coercion, structural forces that perpetuate poverty, government policy,[7] and guest-worker programs,[8] as well as the qualitative conditions of the labor process itself. That is, "degrading conditions" both are a symptom of unfreedom and also "matter in and of themselves"[9] where the "qualitative experience of exploitative work" both results from and contributes to conceptions of unfreedom.[10]

Contrary to the Worm Queen describing herself as "completely free," this brief review of labor freedom suggests the opposite; immigrant worm pickers appear to be working under emblematic conditions of "unfreedom," taking an arduous, muddy, nighttime job in an underground economy to earn a wage for survival. And yet, under these conditions, pickers consistently used the words "freedom" and "autonomy" to describe the primary advantages of worm picking. How can we resolve this contradiction between structural conditions and qualitative experiences of unfreedom and the expressions of freedom by the workers themselves?

Labor geography offers some resolution. Contrary to capital-centric analyses that see workers as an abstract unit of production passively shaped, organized, and exploited by capital, labor geography views workers as active agents in organizing, resisting, and ultimately shaping trajectories of capital accumulation.[11] While initially focused on formalized collective organizing, a plethora of case studies have documented and analyzed diverse forms of "constrained" worker agency that can occur even within some of the most oppressive and exploitative class relations.[12] Intersectional analyses have also opened discussions around how the ontology of work and workers, types of payment, the racialization of labor, and non-waged reproductive labor shape worker agencies.[13]

While labor geography has documented the role of both labor and capital in shaping the built environment and "landscapes,"[14] less attention has been paid to how the materiality of production and its ongoing transformations impact labor regimes and worker agency. Put another way, how do the "unruly," "uncooperative," and "lively characteristics" of nonhuman nature, often documented by resource geography,[15] affect worker agency and expressions of freedom? Recent work has started to make these connections. Elena Baglioni and Liam Campling, for example, note there is a relationship between "ecological indeterminacy" and "labor indeterminacy."[16] Pratik Mishra is explicit in connecting the ecological dynamics of commodity production with labor control regimes through the case of brickmaking in India.[17] Diego Velásquez and Jorge Ayala use the subsumption of nature framework to analyze how the different materialities of wild fisheries and aquaculture influence collective organizing efforts.[18]

Drawing from the subsumption of labor and nature frameworks is extremely helpful, as it explicitly links discussions of un/free labor, worker agency, and how nonhuman nature shapes the production process. It implicates the structural necessity of wage work—that "legal fiction" of "freedom"—while recognizing the variegated strategies of control and (re)design of both labor processes and nature that affect how work is organized, who is in control, and the different logics of accumulation.

Control over the labor, I suggest, is critical to understanding the counterintuitive conception of freedom and worker agency in worm picking. Marx understands how—under the formal subsumption of labor—workers can be forced to work for wages and lose claim over the surplus value created in production while simultaneously controlling their "vitality, freedom, and independence."[19] Whereas real subsumption turns the worker into an appendage of the machine, formal subsumption includes a "multiplicity of forms of wage, non-waged, free and forced" labor forms that coexist under capitalism, often producing a variety of "temporal frictions, asynchronies, and anachronisms."[20] Indeed, under the formal subsumption of labor, there exists a "contradiction between the relation of monetary dependence of the wageworkers . . . and their autonomy in the regulation of the labour process."[21] Seeing the subsumption of labor and nature as a simultaneous process helps us make sense of the plural and contradictory labor forms that arise within capitalist relations.

To be clear, I am not suggesting that expressions of freedom and autonomy characterize all forms of formal subsumption. Quite the opposite. Industries operating in the formal subsumption of labor produce surplus value through the logic of absolute surplus value production. Intensifying production, disciplining labor, and notably extending the length of the working day are the primary means of increasing surplus value. Marx's description of the rural cottage industries essentially operating as subcontractors to the urban factories shows the extent to which laborers, in this case women and children supposedly in control of their own labor and means of production, are still forced to produce around the clock to survive. The contemporary "gig economy" provides an example of how workers can be overworked despite controlling much of their labor and owning equipment. Uber's labor regime is often cited as a contemporary example of formal subsumption, where the initial strategy to work as an independent taxi driver—taking fares at will, using one's own vehicle—suddenly transforms into the cutthroat competition that drives down daily income.[22] The formal subsumption of labor is not a sufficient condition for the expressions of freedom associated with control over one's own labor.

But as we saw above, there is no massive labor pool from which to draw productive pickers. And this is the hinge that ties expressions of freedom to the formal subsumption of labor. There are not enough pickers, and the structure of the worm-picking labor regime means they retain considerable power over their terms of employment. Viewing the subsumption of nature and labor as conjoined suggests that the extent to which capital can control and (re)design nature will shape its ability to control and design the labor process. The ecological conditions of production therefore shape expressions of "freedom" and "autonomy" while firmly rooting such expressions in exploitative capitalist relations.

In this way, worm picking shares several characteristics with contemporary California strawberry picking. Julie Guthman highlights how investment in fumigation has largely underwritten the productivity of strawberries themselves, which up until today has held rising labor costs in check. But this is coming to an end as strawberry growers describe an increasing recalcitrance of both berries and bodies. The dearth of labor has undermined producers' ability to discipline labor through the same means as they used in the past, when one disposable labor force was easily exchanged for another. Now, efficient strawberry pickers are quite conscious of their value and want to ensure they are picking in the most productive fields. They will scout out fields before picking, only work during seasons with higher piece rates, or simply walk off jobs after a few hours in unproductive fields. As a means to attract a dwindling labor force, some producers have switched from strawberry to raspberry production, since the biological characteristics of the latter reduce the physical demands on the workers. Changing the berry changes the working conditions and therefore the availability of labor.[23]

The nightcrawler capitalists, however, have been unable to redesign the earthworm to change the working conditions. Its recalcitrance impedes the real subsumption of nature (and in this case, the real subsumption of labor), and its biological and physiological characteristics throw up barriers to the logic of relative surplus value and even some ways of increasing absolute surplus value. In particular, the working day—or in this case, the working night—cannot be extended. Gains in productivity have already been maximized through manure and no-till agriculture. Increasing profits for the nightcrawler capitalist depends on the spatially extensive logic of absolute surplus value production. They must find more farms and, critically, more labor. All the coolers and crew chiefs need more people with the skill to pick tens of thousands of worms a night. Their profitability depends on it. And the pickers know it.

No picker I spoke with actually enjoys the act of picking worms. But nearly all of them enjoyed the independence it afforded them. Recall how the Worm Queen described worm picking as "completely free:" "It's not like a company where you have a supervisor. You don't have to work during a set time of hours. . . . In a factory, you can't rest when you're tired. This job, you work at your pace. When you're tired, you sleep; hungry, you eat. When you need time because you are ill, just a phone call, and you rest at home. You don't have to talk to your supervisor, explain to him, talk to whoever and blah blah blah."

Out in the fields, pickers were by themselves, without a supervisor telling them what to do or how hard to do it. Suni—a picker who recently stopped to spend more time with her kids at home—said that before she was a mother, "I loved it. . . . Nobody controls you. Nobody tells you what to do." She could work for as long and hard as she wanted, likening the situation to being "self-employed" or an "independent contractor." Like the Worm Queen, Suni noted the lack of supervision and control over the intensity of work: "Nobody watches over you. You feel hungry? You eat. You want to drink? You drink. You feel tired? You can sit down and relax. If you don't want to pick, you can go back to the car and sleep. Nobody tells you what to do. . . . I think, I think a lot of people like that."

Critically, picker expressions of freedom also relate to mobility and their significant power to move between firms. Once viewed as an "opportunistic" act—a strategy of "flight" as opposed to the "fight" of collective organizing[24]—labor geography has increasingly recognized how these "diverse manifestations" of individual acts have broader implications.[25] The ability of an individual picker to change employers seasonally, weekly, or even nightly shapes the structure of worm-picking labor relations. Much of this is reflected in the floating piece rates. Coolers and crew chiefs must raise piece rates to attract pickers, especially when unpredictable weather conditions or surges in demand impact the demand for worms. When the polar vortex extended the 2014 winter well into spring, coolers had difficulty supplying worms to the United States, and the price of worms skyrocketed. Piece rates rose from $25 to $60 per thousand worms. "It's *like* they had a union out there," one cooler told me (emphasis mine), refusing to sell worms until their price was met. A similar rise in piece rates occurred in the spring of 2020 when the COVID-19 pandemic radically increased nightcrawler demand, and rates rose to $65. There were stories of "retired" worm pickers who were over eighty years old who decided the tripling of piece rates might be worth a few nights out in the field. Even when piece rates remain stable at around $30 to $35 per thousand, pickers will only go to fields where they foresee a productive night and will not hesitate to change crew chiefs if their rented fields have persistent low worm densities.

Some pickers had trouble defining who their boss even was, since the pickers often decide who to work for, which nights to work, and how hard they want to work. "The driver pays us, so I guess he's the boss," one older picker guessed. I followed up, "Would the driver ever fire anyone?" She laughed. "No. Even if you didn't pick anything. They don't exclude anyone. If I feel that I can't pick, I just stop working, and he won't fire me."

Pickers, just like the crew chiefs and coolers, will watch the weather to decide whether they feel like showing up. If it's too hot, too dry, or too windy, it may not be worth the time for them in the field. Mai told me, "If I can only pick a few cans I just stay at home. Because you go for so long and can only make a little bit." She would just call the crew chief and say, "Don't pick me up tonight. No problem." She acknowledged this is the biggest difference between picking worms or working at a restaurant or a factory. While Mai had retired, I asked: "If you call a driver today, would they take you?" She responded, "Yes. If I say I want to pick, they'll come pick me up tonight."

Pickers are unafraid to withhold labor or change employers to increase piece rates. When I spoke with the Worm Queen, the spring picking season had just begun. The coolers' warehouses were still filled with "winterized" worms that would soon be shipped out to fill the first surge of demand with the May long weekend. At the time, the piece rate for 1,000 worms was a mere $18. At that rate, the Worm Queen said, "it's not worth it," and she would prefer working at her job picking orders in a warehouse. Even the decision to go to work depends on the worker and their preferences. A fifty-year-old male picker told me the simplicity of his decision to show up for work, or not: "With this job, if you want to go, and you'd like to go, you go. If you're tired, then just don't go."

When piece rates significantly differ between employers, pickers will often move to the cooler or crew chief paying higher rates. But the rate of pay is not the only reason for switching employers. Aside from piece rates and the productivity of the fields, pickers change employers for individual reasons, which, when aggregated, reshape the relationship of labor to nightcrawler capitalists. Nguyen's crew chief didn't let him smoke on the truck, so he found one that would. Mai found her pay was reduced for coffee that the crew chief provided; she changed. Some drivers locked the truck doors during the night, preventing people from stopping work early for a few more hours of sleep. Other crew chiefs might promise high piece rates but then try and renegotiate later. Again, pickers changed. Coolers and crew chiefs are under pressure to rent productive picking fields (higher worm densities) that will increase picker pay (and their own) while piece rates are stable. They also will try to rent fields closer to the homes of the pickers to reduce commuting times. Supplementary benefits are also offered. Cash advances are made, credit is provided, gas money is reimbursed, free transportation is provided, and equipment is loaned. Wages are paid in cash or check at the will of the picker. Whatever small affordances can attract and retain labor, coolers and crew chiefs will increasingly accommodate.

Pickers laughed when I suggested that crew chiefs or coolers might punish them if they missed work. They knew their labor was the direct source of the coolers' profits, and they were not afraid to abstain from working if conditions didn't suit their liking. "It doesn't happen like that in this business," said Mai, referring to various forms of worker discipline. "They need us. . . . If we are successful, then they are successful." She compared this capital/labor relation with factory work and inadvertently touched on power differentials between the formal and real subsumption of labor: "To be frank, the driver needs the picker. When you work in a factory, you need the factory."

Pickers laughed again when I asked whether picking could be remunerated by an hourly wage. They noted that an hourly wage would neither be fair nor worth the effort, as it would compensate productive and unproductive pickers equally. Indeed, there is a form of "popular neoliberalism"[26] in the field, where pickers feel their income is representative of their individualized skill—and this is not something they would give up in exchange for a potentially lower hourly wage. As Suni noted: "Nobody takes care of you. You take care of yourself. . . . If you want to make more money, you work hard. If you don't want to make more money, you're done for the night. You can. Stop. That's why, picking worms, everybody likes it."

"FREEDOM" IN THE CONTEXT OF RESIDENCY STATUS

The lack of labor incentivizes the nightcrawler capitalists to take laborers however they can find them—whether an aging immigrant to Canada, a migrant worker, or even someone working on a visitor's visa. How do these institutional differences in residency status play out in worm picking? How does the materiality of the nightcrawler and its ecology differentially impact pickers based on residency status? If immigrant and migrant laborers do indeed face multiple "layers of vulnerability," how do they navigate these institutional challenges and assert agency, however constrained it may be? How do coolers make use of the Temporary Foreign Worker Program, and is it a solution for the chronic labor shortage?

There are three groups of pickers that differ in their immigration status, each facing different opportunities and challenges in worm picking. Worm pickers who are *permanent residents* or citizens of Canada can obviously enjoy social benefits, such as the education and healthcare that come with such status, as well as political rights that accrue to Canadian citizens. To qualify for some of these benefits, however, pickers need to have a record of employment. That is, they must decide whether to be paid formally by check (and be taxed) or in cash. The former option can be useful, considering that payment by check establishes a record of employment, which may be necessary for various other means-tested benefits, tax credits, or immigration issues. In one case, for example, an immigrant worm picker was not allowed to sponsor his wife to immigrate to Canada because he had no demonstrative evidence of employment or the ability of either of them to support themselves, as required by section 19(1)(b) of the Immigration Act: "The appellant provided no documentary evidence related to this employment."[27]

Resident worm pickers can also qualify for employment insurance (EI) during the winter months and receive 55 percent of their salary, provided they have paid taxes and worked the necessary 420 "insurable hours" and continue to seek employment during the time they are unemployed. Suni, who came to Canada in 2004, picked worms in the spring, summer, and fall, and then claimed EI during the winter months. She estimated she received about $450 every two weeks, noting, "That's enough for winter." This also allowed her to take adult education and business classes. Coolers' employees who work in the warehouses packing and sorting worms will also claim these insurance benefits.

And yet, most resident pickers do not choose the aboveground route. Coolers estimate that between 75 and 90 percent of pickers are paid in cash and avoid taxes. Nguyen, the worm picker/stock trader, used to claim EI benefits in the past, but eventually, he decided it wasn't worth the hassle. He didn't like having to document how he was looking for work in the off-season and preferred to travel back to Vietnam in the winter. For the last several years, he has always been paid in cash. "When I'm finished [for the season] I just travel. I don't have to report. Don't have to ask for permission." He knew he was at risk of being caught avoiding taxes, but he didn't seem too fearful (and I have found no court records of the CRA charging pickers for tax evasion, as opposed to crew chiefs and coolers). As residents of Canada, therefore, immigrant worm pickers are legally free to work for whomever they choose, largely under the labor arrangements that they choose (when to work, how much to work, transport to and from work, paid in cash or check, etc.). Any type of grievance they have with the employer can be rectified by finding a new cooler or crew chief, with little fear of legal repercussions. Their status as permanent residents or citizens grants them access to all of Canada's social benefits, which they are free to utilize based on their individual circumstances.

Some coolers and pickers noted that an increasing number of pickers may not possess legal work visas. These *non-status pickers* include people picking worms on a tourist visa, people who have overstayed their visas, and those on a temporary foreign worker visa unrelated to worm picking (that is, they are illegally worm picking outside of their formally recognized work). It is not rare for people from Vietnam, Greece, Turkey, or, as I heard, "one Polish guy" to arrive in Canada for the summer and pick worms on a tourist visa. At the end of the picking season, they simply return home—a sort of informal temporary foreign worker program that relies on familial connections to the worm-picking industry. The risk for these pickers is serious, involving deportation and prevention from re-entering Canada. Other non-status migrants include those who have overstayed their visas. I spoke with two Thai translators who had been employed by the Canada Borders and Customs Agency (CBCA) during their "raids" of temporary foreign workers who had either overstayed their visa or had been employed by one business but were then found picking worms at night for cash—contravening their single-employer contracts and tax obligations. I also connected with two people who came to Canada as temporary foreign workers but decided to run away from their employers and stay in Canada. "I came to earn money," said one picker, "so you do whatever job earns money."

Obviously, these pickers face additional layers of vulnerability compared to their worm-picking counterparts with resident status/citizenship. Non-status pickers cannot be paid by check, nor can they claim social benefits that accrue through either legal residency or formal employment. Any interaction with public services (although technically possible) is extremely risky. Indeed, immigration status in Canada is a "major factor affecting such a person's ability to seek, and experience of, healthcare services,"[28] as well as other educational or workplace safety benefits.[29] Such pickers are also extremely vulnerable to injuries caused on-site or in traffic, as there would be no avenues for support that wouldn't also end in their deportation. Nor did those I spoke to ever expect to return home. This was the biggest difference they saw between themselves and the pickers with resident or citizenship status: their undocumented migrant status effectively bars them from returning to their home country to visit family.

Given these precarious constraints, I was surprised, however, that these pickers continued to cite "freedom" and "autonomy" as the primary benefits of worm picking. The difference between residency status, however, meant that their conceptions of freedom were relegated to the working conditions out on the field. The materiality of nightcrawler production allows them to not have a "boss" watching over their shoulders, and they equally appreciate being paid for the effort they put in. Nor were they bound to contracts, so they could seek other opportunities that might arise. It also offered the possibility to make more money than other positions they have had in Canada, such as packing vegetables in a factory; over the past two years, one of the pickers claimed she had earned between $30,000 and $40,000 picking between 100 and 120 nights a year—at least three to four times the amount she would make working a similar number of days at a minimum-wage job.

They also expressed an additional benefit of worm picking compared to other low-skilled jobs—one that was based on their precarious and undocumented status. In a context where they needed to remain invisible and "off the books," the specific materiality of the production process of worm picking provides extraordinary levels of anonymity. The bait industry's underground and largely informal character allows them to work alongside permanent residents and citizens on relatively equal footing, being paid in cash with little physical evidence of employment. Work occurs at night in the dark, on rural agricultural land, and work sites change on a nightly or weekly basis. For resident worm pickers like Nguyen, this allows them to avoid taxes. For non-status workers, the clandestine nature of the work gave them the best chance to stay below the CRA and CBCA radar while making tens of thousands of tax-free dollars in Canada.

Both resident and non-status worm pickers are fully aware of their value and do not appear afraid of any direct forms of discipline for actions such as changing crew chiefs for better pay, more productive fields, or other benefits. They can enter and exit the industry based on their evolving individual circumstances, be it other job opportunities, health issues, child-rearing, or family care. The freedom of worm picking makes it difficult for individual coolers and crew chiefs to retain worm pickers at a cheap price. To paraphrase Don Mitchell, coolers need the critical commodity called *labor*, but all they can find is *people* ready to assert their autonomy over the worm-picking labor process. Is there a way coolers and crew chiefs can revert *people* back into *labor*?[30]

TEMPORARY FOREIGN WORM PICKERS

Gordon, the cooler with the hairdressing spokeswoman, landed in Bangkok, Thailand, and immediately began knocking on government-building doors hoping to find someone to help him find potential worm pickers. After being shuffled around various government agencies, he finally secured a meeting with an official in the Thai Ministry of Labor, who told Gordon to meet him at a parking lot at a specific date and time. Gordon obliged. "Oh my gosh, when I arrived there, you know, at the parking lot, it's fucking bigger than my property. A thousand people, they show fucking up. They [the officials] said, 'Pick anyone you want.'" How did Gordon choose his fifty workers? He didn't care if they were men, women, married, single, divorced, or had kids. He only required healthy twenty-five- to thirty-five-year-olds who could pick worms fast. "How are you going to test it?," he remembered the Thai official asking him. Over the next several days, Gordon went to a park in Bangkok to conduct his "tests." He had exchanged $50 for small Thai baht coins and began walking around the park, throwing the "pennies" in each and every direction, "Shew, shew, shew! Spread all the pennies through the park."

Gordon told the potential pickers, "I'll take whoever picks the most pennies in ten minutes." Some, he said, came back with a mere two or three pennies, others with fifty. He dutifully recorded their names and totals. After several days, he narrowed his choice down to fifty pickers. Gordon submitted all the necessary paperwork, and the Thai worm pickers arrived the next May, picked worms for six months, and then boarded a plane back to Thailand.

Gordon's Thai pickers were not *immigrants* with residency or citizenship status, flexing their agency in the worm industry, but *migrants* employed through the "agricultural stream" of Canada's Temporary Foreign Worker

Program (TFWP)—a type of guest-worker program, which allows employers to hire foreign workers for restricted periods of time. Through this program, Gordon was able to reassert control over the labor force in a way that he could not with immigrant worm pickers. These Thai pickers were legally required to work for him, *and only him*. They had to pick worms on the fields *he* rented, transported by *his* crew chief on the nights *he* wanted, picking a quota of worms determined by *him* (which at the time, was 800 per hour). "If they don't pick the minimum, go home. . . . I just call the labor department and say I got this person I can't keep because I can't afford it." Such Temporary Foreign Workers (TFWs) are institutionally immobile and attached to a single employer, who can select the labor of their choosing—gender, age, nationality, productivity—and insert them into a production process only when they are needed. At the end of their contract, they must return to their home country, where they must wait and see if they will be rehired in the next production season. It is obviously impossible to fully transform people into objective factors of production, like a piece of machinery. But programs like the TFWP certainly try.

Proponents of the program suggest temporary foreign workers fill urgent labor demands in a variety of sectors and characterize the program as a "win-win" for both the employer desperately seeking labor and the workers, who tend to come from poorer countries with high rates of unemployment and low wages.[31] Workers purportedly enter contracts voluntarily and are provided health insurance during their time in Canada. Their flights are covered by the employer and often their lodging is subsidized. Proponents also rebuke the idea that these migrant workers are cheap labor, since they are legally required to make the same amount of money as a resident—either minimum wage or the prevailing rate of the industry. As these low-skill TFWs tend to come from less industrialized countries, their Canadian wages are, at least nominally, significantly higher than in their home countries. Hence, the apparent win-win championed by advocates: employers get labor, and the labor gets wages unattainable in their home countries.

Critics, however, describe the TFWs as facing multiple "layers of vulnerability"[32] where low-skilled, physically demanding labor is outsourced to racialized populations from developing countries who have little hope of permanent residency. Often labor is procured through "recruiters" in the source countries (although the practice is technically illegal) who charge exorbitant fees, frequently leaving the worker in significant debt before beginning the work term.[33] And while housing is often provided by the employer, this often takes the form of "bunkhouses" on-site, thereby allowing the employer to be both boss and landlord, capable of monitoring both working and nonworking hours. The most trenchant critique of the program, however, is directed at the single-employer contracts that stipulate TFWs can only work for the employer that sponsored their visa. They cannot quit, change jobs, or move to other areas without the risk of immediate deportation. The inherent "immobility" of the program[34] ensures that there is no practical mechanism for migrant workers to change employers and seek increased wages or better working and living conditions. Their reliance on the single-employer contract also discourages them from submitting formal workplace grievances about issues such as unsafe working conditions, wage theft, nonpayment, expected quotas, the inability to access proper healthcare, or workers' compensation. Many scholars criticize the single-employer contracts as a form of "bonded labour,"[35] "indentured labour,"[36] or working under the condition of "unfreedom."[37] Canada's United Farm and Commercial Workers has called the program "Canada's dirty little secret."[38]

Based on newspaper records, the first cooler to apply for and receive TFWs for the explicit purpose of worm picking was National Bait in 2004. It did not go well. National Bait initially employed forty Mexican workers through an earlier version of the TFWP. Shortly after the migrant workers arrived, they claimed their wages were reduced to $8 an hour, down from the promised $10. They also felt the 5,000-worm nightly quota was unreasonable and complained they lacked water and latrines on the picking fields.[39] National Bait, on the other hand, claimed the workers did not have the appropriate backgrounds and were not accustomed to such physically demanding labor, blaming the Mexican consulate for their poor choice of workers: "We look for farm workers and we got engineers, secretaries, a mayor's assistant." Others were storekeepers and clerks. "They didn't want to work that hard. I just started to send them back. We need people who want to work."[40] Eight underperforming employees were sent back after two weeks on the job. Another eighteen would be sent home throughout the season.[41]

In this way, the TFWP offers coolers the chance to control labor in ways they cannot with other resident pickers. And yet, not just any TFW will do. Gordon's ad hoc field trial in a Bangkok park selected for productivity. In some ways, they need to be *more* productive than workers already resident in Toronto, considering the cooler must recoup the $1,000 Labour Market Impact Assessment (LMIA) processing fee (see below), the flights, and the costs of providing affordable accommodations. One cooler who has received workers in the past estimates he has spent close to $4,500 on each temporary picker for such costs.[42] For this expense, he wants to find the most productive pickers and maximize their time in the fields.

The ability to rehire (or not) the same pickers year after year also allows employers to filter out unproductive workers and select for "productive" characteristics such as nationality, gender, and age. Gordon wanted certain ages. Another cooler only wanted men as a way to eliminate what they called "funny business" happening after work hours. Through trial and error, coolers have also cycled through different nationalities, trying to find a population that can not only tolerate the demands of worm picking but do so productively and without complaint. They've experimented with employing Jamaican, Filipino, Mexican, and Vietnamese migrant laborers, with each deemed problematic for one reason or another.

Surprisingly, piece rates for TFWs fluctuate over the season depending on the market demand for worms, as opposed to the fixed wage stipulated in their TFW contracts. Why would coolers increase the rate of pay for TFWs? The problem, as one cooler told me, is that "pickers talk" and will quickly find out who is getting paid at what rate. Another cooler knows he could probably get away with paying lower wages to TFWs but also doesn't want to risk aggravating the productive migrant workers he has selectively chosen over the years. When he finds good pickers, he said, "I want them coming back here!" However, though rates of pay for TFWs did fluctuate over the season, they were consistently lower than those for domestic pickers. In 2020, when the COVID-19 pandemic sent the piece rate up to a peak of $60 for residents and non-status worm pickers, the TFWs were making $40 per thousand. When prices stabilized later in the season, they were making $20 per thousand, compared to the market rate of $25. The TFWs were clear that the money was relatively good compared to both their home country and other jobs they had either done or heard about in Canada. Most of them made between $1,500 and $2,000 per month from May to October—a large increase over the monthly minimum wage of approximately $252 in Thailand. Each of the interviewees was also keen to keep their position and did not want to jeopardize their position by accepting other work in cash, overstaying their visa, or trying to find another employer.

TFW pickers still cited "freedom" and "autonomy" as the primary advantages of worm picking, but expressions were much more constrained. They did not have the freedom of mobility, nor could they even take a second job if they wanted. They could not walk off the field after a couple of hours, nor could they travel to a more productive field or find a preferable crew chief. Instead, their expression of freedom and autonomy related only to their work in the field.

The TFWs I spoke with didn't identify structural differences between themselves and other resident worm pickers but acknowledged they wouldn't have any way to really know. Like the other pickers, they enjoyed not having a boss monitoring their work, and they too liked how their compensation was determined by the amount of effort they exerted. They shared the same physical concerns as other pickers, such as back and knee pains, by far the biggest complaints from all pickers, but they said they didn't feel pressure to go to the fields if they needed a day of rest or were feeling ill. And yet, none of the TFWs I interviewed took a day off; they were losing money each day they spent not picking. Indeed, one of their biggest complaints about the TFWP is that they were not working enough!

The TFWs I spoke with expressed this frustration not at their employer but at the worms. As one of them said, "You can't work every day. You have to depend on the worms." They estimated they worked between three and five nights per week. Sometimes in the summer they were not working at all, and simply stayed in the housing provided by the cooler. This was frustrating for them, as they came to Canada to make money, but the materiality of nightcrawler production often prevented them from doing so. Their single-employer contracts also prevented them from working more hours for other potential employers, whether in the worm picking business or not. As a result, when worm picking conditions aligned and the worms were out, they were ready to be as productive as possible. I asked whether they feared being deported for not reaching their quota. True, they did have a quota, but the pickers I spoke with never had trouble meeting it, usually picking between 10,000 and 15,000 worms on a picking night. They had a fixed time in Canada, and they wanted to make the most of it.

At first glance, the TFWP offers a direct solution for coolers perpetually struggling to find willing bodies to pick worms. But there is a problem; the peculiarities of the bait industry—shaped by the formally subsumed nightcrawler—do not easily translate to the bureaucratic government program. The limited public knowledge of the industry, combined with the piece-rate wage structure and ephemeral "production sites" makes for a convoluted application process full of inconsistent oddities that other industries, like an organic farm, largely avoid.

As stated earlier, companies seeking to employ TFWs must obtain a favorable Labour Market Impact Assessment that adequately demonstrates the lack of locally available labor. Already, the bait industry is at a disadvantage. The industry as a whole remains largely unknown, unregulated, and lacks a lobbying or marketing association; most people do not even know it exists, let alone the amount of labor required, the pay structure, the hours, or the working conditions. Gordon hasn't received a positive LMIA in several years, and Apollo Baits, who had received TFWs in the past, now finds himself rejected with each costly LMIA: "Every time I argue with them, I tell them, you know, I'll pick you up. I'll take you out and tell me if you want to do this business or if you think your son or daughter should take this as a career move. I think they get offended." Compounding the bureaucratic problems, the LMIAs must be completed using hourly wages and fixed workplace addresses. But pickers are paid in piece rates that fluctuate through season, working on different pieces of the land. Trying to document all these complications raises red flags from the labor-market assessors. When coolers submit their LMIA, four to six months prior to the start of the worm-picking season, they may not yet know where the TFWs will be working. As one cooler noted, this results in a disorganized LMIA always in need of clarification: "They [the LMIA officials] want the addresses of each field, where they're picking. Well, we rotate, so, oh my goodness! They want addresses, we give them, then they want another," he shook his head as his voice trailed off, then concluded, ". . . the government."

Many frustrated coolers, tired of throwing away tens of thousands of dollars at failed LMIAs, have given up on the program. "It's really complicated," Apollo Baits said. "And that's part of the stupidity of the government. . . . If they really wanted to, they would realize there is an incongruity between how the business operates and how the government thinks a business operates."

Until very recently, only one cooler has managed to consistently get TFWs for worm picking over the past decade. Other coolers suggested this specific cooler has "connections" in the government who approve his LMIA and purposely scuttle theirs. Parga Baits sums up the general attitude among the competing coolers: "You give to [cooler name], but not to me. Why? Something is going on for sure." A representative with Employment and Social Development Canada, the governmental branch that conducts LMIAs, assured me, however, that the low-level, relatively mundane bureaucratic processes would not allow for such malfeasance.

Even if these coolers are frustrated with their own LMIA decisions, they are grudgingly content that, at the very least, *someone* is securing external labor through the TFWP. In 2014, for example, when one cooler was denied TFWs, the ramifications of fifty fewer bodies picking worms spread through the industry. Relationships between coolers and crew chiefs, and crew chiefs and pickers, fell apart, and agreements were broken as pickers sought to sell their labor to the highest bidder. I asked Apollo Baits if fifty or sixty workers really make that much of a difference. "Oh, fuck yeah. It's all labor. . . . That's 50 to 100 laborers we don't have to compete over." Indeed, a napkin calculation of fifty pickers averaging a mere 10,000 worms a night over 100 nights would represent 50 million worms over the course of a season. Any additional TFWs for one cooler is welcome news to all the others.

· · ·

The formal subsumption of worm-picking labor prevents nightcrawler capitalists from asserting full control over their workers, who are free to change crew chiefs, set their own hours, and jump at other opportunities when they arise. This enables a surprising expression of "freedom" that was largely shared by each of the pickers I spoke with. Such expressions of freedom, however, are not identical. We can see how "im/migrants" are not a homogenous group always subject to the same vulnerabilities but actively navigate these institutional benefits and constraints based on their circumstances. Immigrant worm pickers have much more control over their labor conditions because of their legal rights in Canada, choosing how, when, and where to pick worms. Non-status immigrants are afforded few legal rights and are under constant threat of deportation. The "benefit" to worm picking for non-status immigrants is the ability to make money in a relatively invisible industry with little paper trail or fixed locations. In contrast, current TFWs do not have the freedom to change employers for better pay or working conditions and must work on the sites and nights of the cooler's choosing. Their biggest frustration is not being forced to reach a necessary quota of worms—which they can easily do—but rather that they are unable to maximize their earnings in Canada because their work is structured by the ecological conditions of worm habitat and the biology and behavior of the nightcrawlers. For coolers, the TFWP initially appears as a useful mechanism to increase labor, but again, the specificity of nightcrawler production stuck in formal subsumption creates an extremely convoluted bureaucratic process to secure TFWs.

Every cooler I spoke with cited the lack of labor as one of the primary threats to the future of the industry. Worm pickers are an aging demographic, and the TFWP has yet to fill the shortage of willing bodies picking worms for pennies apiece. But this is not the only threat to the industry. The share of surplus value going to the labor is shaped by the movement of other elements in the nightcrawler assemblage. The problem of labor could be soothed, for example, if farmers didn't charge such high rents, if climate change didn't reduce the number of optimal nights, or if more recreational anglers were willing to pay more for nightcrawlers. Over the past decade, each of these elements has been pulling in directions that threaten the future accumulation of the nightcrawler capitalists. I turn to these future challenges and explain how a new and unexpected element in the nightcrawler assemblage provided a brief reprieve to an otherwise bleak-looking future.

SEVEN

How a Bait Assemblage Falls Apart

IN THE EARLY SPRING OF 2020, one of the largest regional wholesalers in the United States attempted to discipline their Canadian nightcrawler suppliers by setting the price at $60 US per thousand. They made it known they would not negotiate, and the Canadian coolers could either sell at this set price "and not a penny more" or find buyers elsewhere. But things quickly changed. In March of 2020, governments around the world began declaring states of emergency as COVID-19 cases starts rising exponentially. Schools, libraries, churches, and "non-essential" businesses such as gyms, theaters, and restaurants were ordered closed. Bait-and-tackle shops were not spared from the shutdown either, and coolers became concerned their wintered-over worms might not have any buyers. Lost sales piled up week after week into May. Initially, as DCA Baits said, "We were all worried."

But as the weather started warming up, DCA Baits noticed the smaller convenience stores and gas stations—outlets that would stock 10 to 20 dozen worms in their mini fridges—kept wanting more nightcrawlers. When the larger bait shops reopened at the start of summer, demand for worms exploded. The large American wholesaler, which had said it would not negotiate, was forced to pay over $100 US for a thousand worms. As it turned out, recreational fishing became a very popular activity during the COVID era, known as "social fishtancing." And these were not just the same old fishers of years past. One major retailer in Ontario noted that by May and June there was a disproportionate increase in orders for entry-level and beginner equipment, a trend that was documented in the United States and England as well. The retailer was cautiously optimistic these anglers would continue their new hobby as sales remained strong in 2021: "We have retained many of the new anglers and those that have rediscovered it." And we know what bait new anglers demand.

Prior to the COVID-era surge in fishing participation, the nightcrawler industry was arguably in decline, facing serious challenges on several fronts. The ties binding the assemblage of human and nonhuman actors and processes that had aligned between the 1950s and 1980s were beginning to fray. The ability to capture surplus value—the thread holding the actors and processes together—was coming up against what seemed to be insurmountable challenges. For some of the coolers, a reckoning in the industry was coming or had already begun. Apollo Baits told me he intends to ride it out as long as he can before he closes up shop. "There is no exit strategy," he told me. No one is going to buy his business.

In this chapter, I revisit some of the key elements in the nightcrawler assemblage and how they relate to one another in ways that constrain or potentially shift to facilitate accumulation in the industry. These include "scarce" labor, land, customers (aside from the recent COVID bump), and picking nights due to climate change. The scare quotes are intentional. Political ecology has long noted that "scarcity" is something produced and relative, as opposed to an objective accounting of dwindling finite resources. In the nightcrawler assemblage, we can see how these scarcities of land, labor, consumer demand, and prime picking nights are interrelated, with unexpected perturbations in one element reverberating through the whole. A changing climate, for example, affects precipitation, which affects nightcrawler behavior, which affects the number of picking days, which affects piece-rate wages, which affects whether pickers show up for work, and so on. At the same time, there are unforeseen surprises that could feasibly realign the elements and make surplus value production much easier. COVID-19 unexpectedly increased fishing participation rates, which increased demand, which increased wholesale prices, which increased piece rates and attracted more laborers while rents remained constant.

The point I want to stress is how nightcrawler capital (personified in nightcrawler capitalists) cannot easily control the multiplicity of human and nonhuman actors and processes necessary to produce surplus value; coolers and crew chiefs are perpetually experimenting with the people, places, conditions, and things available to them. Sometimes new elements are enrolled in the assemblage; sometimes others are thrown out, adapted, or realigned to allow continued capital accumulation. If capital fails to accumulate, the assemblage will fray. The question for the bait industry is whether the current and new elements can be successfully realigned for a profitable future. Despite the COVID-19 bump in recreational anglers, the prognosis is grim.

SCARCE LABOR

While the TFWP offers employers access to a global labor supply, the idiosyncratic organization of nightcrawler production has hampered the ability of nightcrawler capitalists to utilize the program. In this sense, we can see how the term "labor shortage" exists only in relation to other socio-ecological, economic, and political forces. It is not that Canadian residents don't want to pick worms for money; the problem is they cannot do so productively at the offered wages. But what if the piece rate went up? Could labor be found? In short, yes. When the market demand exploded for worms and regional buyers in the States were willing to pay more and more for the scarce worms, piece-rate wages rose to unprecedented levels—up to $60 per 1,000—and labor could be readily found. One cooler reported an eighty-two-year-old Vietnamese man who came out of "retirement" to pick for the 2020 season. Most of the coolers told me they would readily pay higher wages if they could—not out of altruistic intentions, but rather to attract the more productive laborers away from their competitors. Aside from the brief flexibility the COVID bump offered them to increase prices, however, coolers describe themselves as squeezed by their buyers on one side and the powerful dairy-farming earthworm rentiers charging astronomical rents on the other. Is there any way to reduce the slice of the surplus going to the rentiers?

SCARCE LAND

Like the concept of a "labor shortage," "scarce land" is also relational. Despite the high rents dairy farmers charge crew chiefs and coolers, nightcrawler-dense land is *not* in short supply in any absolute sense. As I described above, every dairy farmer growing alfalfa and spreading manure is an inadvertent nightcrawler farmer. According to OMAFRA, there are 1,161 dairy farms in the five counties (Middlesex, Perth, Waterloo, Wellington, and Oxford) where most worm picking occurs.[1] Imagine for a moment each of these dairy farmers leased a mere 20 acres (1,161 × 20 = 23,220 acres). Multiply this by a measly yield of 100,000 worms per acre per year, and we have a total of over 23 billion worms—enough to satisfy global demand at least forty times over. The problem for nightcrawler capitalists is thus not a lack of land or worms, but rather a lack of farmers willing to rent their land. Many dairy farmers distrust actors in the nightcrawler industry and would rather not get involved with the headaches and squabbles they have experienced or witnessed in the past. Some remain concerned about the impact of worm picking on their land, and although they recognize that it may not be especially harmful, they also assume it can't help their soil either.

For nightcrawler capitalists, the land problem is, at least theoretically, one of the easier challenges to manage. As we've seen, rents remain high for two main reasons. First, dairy farming is protected through Canada's supply management system. The guaranteed prices for milk and quota limits stabilize income for dairy farmers. As a result, there is little necessity or pressure for farmers to engage with an unregulated underground industry or have a bunch of people duck-walking around their property in the middle of the night. That being said, supply management remains persistently controversial in Canadian politics. Calls to dismantle the system—typically from conservative politicians, scholars, and institutes—persist and the Canadian government has historically offered pieces of the Canadian dairy market as a negotiating chip in free-trade agreements (having given up 10 percent of the market in the last few years alone). Increased economic uncertainty for dairy farmers has the potential to incentivize other revenue-generating activities, such as worm picking. I am not convinced that all farmers in the region know exactly the prices they could command or the fact that no additional expense is needed.

The challenge for the coolers and crew chiefs is to soothe farmers' fears about engaging in the "fly-by-night" or "underground" character of the industry. Phone calls at 3:00 a.m., garbage left in the fields, unfulfilled promises of payment, bags of untraceable cash, police lights, murmurs about drug and motorcycle gangs, disconnected phone numbers, and persistent renegotiations, combined with a general distrust of "foreigners" speaking in broken English, all discourage farmers from further integrating with the industry. In contrast, formalizing contracts, paying up front, and enforcing explicit rules on the field that prevent garbage, conflicts, and visits from the police would allay many of the farmers' concerns and inconveniences. Farmers on good terms with coolers and crew chiefs have relatively few problems engaging in the industry and base their rental price largely on what they have heard others are getting.

Also, few actors in the industry have heard about the potential ecological benefits of renting land. Indeed, promoting the counterintuitive (and anecdotal) observation that renting land to worm pickers could improve soil health by leaving the land fallow, kick-starting juvenile *L. terrestris* population growth, and reducing compaction, all while being paid more money than growing any other cash crop, could encourage more farmers to see the benefits of engaging in the industry. The challenge is finding farmers with more land than they need for their own dairy operation. However, if coolers and crew chiefs are already willing to pay for manure application, renting fields from no-till cash croppers would also be possible. This would increase the complexity of the rental agreements, as crew chiefs and coolers would have to find fields near dairy operations that do not have enough land to spread their manure nutrients. They would have to engage in one contract for the field, another contract for the manure, and ensure it is spread at the right times throughout the year. I suspect, however, that cash-croppers would be more than pleased to receive income, improve the nutrients of their fields, and eliminate expenses. The ad hoc nature of such an arrangement would no doubt be cumbersome for the crew chiefs and coolers, but it does demonstrate the variety of ways production could be reorganized to incorporate more land into the picking industry.

Increasing the number of farmers willing to negotiate contracts would increase the supply of land and, theoretically, create the conditions for lower rental costs. Based on interviews, if the hassles cited above were properly managed, farmers could probably be convinced to lower rents from over $1,000 an acre to approximately $500 to $600 an acre per year. This would still be double or triple the regional land rental rates and much more profitable than growing corn, soy, or wheat. The decision for land-short farmers engaging in last-cut arrangements—that is, those who do not have enough land to rent for a full year of worm picking—is also much more flexible in negotiations, considering they need not change any farming practices nor have any additional expense.

The other option to address scarce land, and one which I can only hint at, is avoiding farmer fields altogether and cultivating the nightcrawler in a closed system. This has been the dream of nightcrawler producers for decades—the real subsumption of nightcrawlers. As noted in chapter 2, *L. terrestris* are notoriously difficult to cultivate. This statement is true, but it comes with a massive caveat. They cannot be cultivated *in an economically feasible way*. This could, at least in theory, change. In the 1980s, a dozen worms cost about a dollar. As the price of nightcrawlers rises, however, the economic case for nightcrawler cultivation becomes stronger. Early in my research, I talked with one individual who is experimenting with producing nightcrawlers; this person claimed they are close to a marketable product, especially considering the increasing cost of land and labor. Another cooler had a feasible idea he shared to massively increase the productivity of picking but requested I not make the strategy known. Either option would likely require fishermen to pay more for worms than the already high price of $4 to $7 US per dozen. Is there a market for such high-priced live bait? The elasticity of nightcrawler prices is largely unknown, and fishermen I spoke with suggest $6 would be way too high. One retailer raised his prices in 2019 to $4 per dozen and was slightly surprised people were still buying. When I started calling bait-and-tackle shops, the average was about $4. Yet one North Carolina store sold a dozen nightcrawlers for $6.99 US! Much of this depends on demand, which saw an unexpected surge in 2020 thanks to a novel coronavirus.

SCARCE CUSTOMERS

Up until the COVID boost in demand, coolers, regional wholesalers, and the retailers I spoke with all cited stagnant demand. The halcyon days of the nightcrawler industry of the 1980s, when prices were determined in the colorful early-morning auctions on Pape Street in Toronto, have long since passed. As one large cooler suggested, "the market is fully mature," with little room for future growth. Apollo Baits explained his customer dynamics well: "We're not finding new customers; we're just taking other people's customers, and they're taking our own. Just recycling sales. There are no new customers." The younger generation of coolers are biding their time, waiting for the older generation, particularly those immigrants who started in the 1970s, to finally retire so they can pounce on their customers.

The rapid growth of recreational angling, as documented in the early Fishing and Hunting Reports, plateaued in 1985, with 39 million freshwater anglers. By the following survey in 1991, this number had declined to 31 million, which has largely remained stagnant ever since, likely counting the same cohort of recreational anglers with each five-year survey (table 3). Why aren't there new anglers?

Coolers mostly blame shifting recreational norms. Ben explained to me how kids "used to go out with their grandfathers and go fishing. Now they sit in front of a screen." Television, video games, the internet, and iPads were all cited as usual culprits. One large retailer in Ontario had noticed the downward trend across the entirety of the angling industry, estimating that most of his customers started fishing between 1960 and 1980—most likely the same individuals captured in the five-year fishing participation surveys. He told me: "The market is shrinking. . . . It's something we've been talking

TABLE 3 Recent trends in US recreational freshwater fishing rates

Survey year	Number of freshwater anglers (in millions)	Average number of days an individual spent freshwater fishing
1980	36.4	20
1985	39	21
1991	31	14.2
1997	29.7	17.3
2001	28.4	16.4
2006	25.4	17
2011	27.5	16.5
2016	30	16.5

SOURCE: Data from the US Fish and Wildlife Service (1980, 1985, 1991, 1997, 2001, 2006, 2011, 2016).

about for over 10 years. . . . A lot of traditional pastimes are experiencing a decline amongst younger populations who have so many other options. Online gaming is probably the biggest one. Guys are better at playing online hockey than actually playing hockey."

Coolers, for their part, have tried to diversify their market and are currently supplying the famous "Canadian nightcrawler" back to its native European homeland. Some of the midsized and larger coolers suggest 30 to 50 percent of their sales are now airfreighted to Europe, where the popularity of nightcrawlers as recreational fish bait is increasing, according to one English retailer.

It is worth mentioning three other dynamics that further constrain the market opportunities for Canadian nightcrawlers: live bait regulation, the increasing use of the European nightcrawler (*Dendrobaena veneta*) as bait, and artificial bait. I have constantly referred to nightcrawlers as the most popular "live bait" for recreational freshwater anglers in the colloquial sense, as opposed to the regulatory sense. According to the Ontario Ministry of Natural Resources and Forestry—alongside equivalent ministries across the provinces and US states—the definition of "live bait" excludes worms and refers to baitfish (like minnows) and leeches, often considered invasive in aquatic habitats. Commercial "live bait" companies—those who sell minnows, leeches, grubs, and such—require a "bait license" *unless* the company's sole activity is buying, selling, and transporting worms. Worms have received this free pass because, first, they do not live in water and pose little threat to aquatic ecosystems, and second, they have historically been viewed as a *beneficial* macro-organism in soils. Wouldn't throwing unused nightcrawlers on the ground be good for the soil?

Whether nightcrawlers are considered "good" for the soil depends on the soil and ecosystem in question. Throwing nightcrawlers into temperate zone agriculture would be welcomed by farmers. Throwing them in a backyard garden might be helpful (though this should not be done during the day, as I once did with 1,000 worms. Word of this free lunch spread fast to the robin community, and my backyard became a scene from Hitchcock's *The Birds*). What about throwing the unused worms across the forest floor next to the fishing lake? This must be beneficial as well, right?

No, it is not.

In the late 1990s, forest ecologists started asking why there was so little understory in the Minnesota forests they were studying. After some investigation, they realized that nightcrawlers were consuming the entire leaf litter on the forest floor, a severe and intractable problem for ecosystems that evolved for the past 12,000 years without the presence of earthworms. Without a thick layer of duff of slowly decaying organic matter, native plant seeds could not germinate. So where did these nightcrawlers come from? "Think about it," one of the researchers, Cindy Hale, says, "You've been at the lake fishing all day. You still have a few bait worms left when you're finished. What do you do with them?" Dump them on the forest floor.[2] Later studies confirmed this hypothesis, concluding recreational fishing is the "major vector" for introducing exotic earthworms into forest ecosystems.[3] Such findings have been gaining attention in the popular press, with articles in local newspapers as well as the *New York Times* and on the Canadian Broadcasting Company, with calls for increased bait regulation that includes restrictions on the use of worms. Many fishing jurisdictions have moved beyond implementing "live bait" restrictions to now state "artificial bait only," ensuring no living bait, earthworms included, can be used. Several states, like Minnesota, specifically prohibit releasing worms anywhere in the state.[4] In other jurisdictions, such as in Algonquin Park, Ontario, anglers are "encouraged to dispose of unused worms in garbage containers (not on the ground)."[5]

The United States Department of Agriculture's Animal and Plant Health Inspection Service has also become more stringent around soil pathogens crossing the border. In June 2022, they suddenly required US earthworm importers to obtain a Plant Production and Quarantine 526 Permit by July 1, 2022. As earthworms may contain plant and animal pathogens, the permit ensured imported earthworms were not to be transported in soil nor have soil in their digestive tracks. Any imported worms must be placed on a "cleansing diet that is free of any materials that may contain plant or animal pathogens."[6] The diet requirements were not usually a problem for the coolers, as they tend to use peat moss as bedding and feed, but the new permit would require them to hold the worms for at least fifteen days in the cooler. More problematic was that the cooler's US clients were all denied these permits in June. One cooler told me the USDA required "more information and clarification" about the process of stocking worms. The picking industry almost collapsed. US clients were trying to buy their entire yearly supply before the July 1 deadline, and crew chiefs and coolers told their pickers they would have to stop working. After this initial shock, on June 10, the USDA extended the deadline to October 1. One cooler was confident this would give him (and likely others) enough time to ensure their US client's permit approvals.

Second, while Canadian nightcrawlers remain the most popular bait worm—long dominating their closest rival, the red wiggler—there is increasing competition coming from a new foe, *Dendrobaena veneta*. Also known as *Eisenia hortensis* or more commonly as the European nightcrawler, *D. veneta* is much larger than the red wiggler. Though it cannot rival the size of the Canadian nightcrawler, it does have one significant economic advantage—it is an epigeic worm that can be commercially cultivated in a closed system. As noted by a retailer in England in chapter 3, it is an "industrial worm" that can be easily produced for either bait or vermicompost. Vermiculturalists in the Netherlands appear to be particularly adept at *D. veneta* production and dominate the global trade for this worm, which has slowly been appearing in bait shops marketed as "Big Red Worms." Tracking the US imports of "live worms" from Canada and the Netherlands is particularly illuminating. Between 2003 and 2020, the total value of American imports of Canadian nightcrawlers varied between $24 and $44 million US per year, including the 2020 COVID-19 boost. Over the same period, the value of American imports of worms from the Netherlands increased steadily from about $3 million to $13 million US, with a massive jump to $43 million US in 2020, effectively equaling the value of Canadian nightcrawler imports.[7] The COVID bounce certainly helped the Dutch *D. veneta* producers as well, providing another worm option for the new wave of recreational anglers.

Third, the artificial-bait market continues to expand, offering anglers new live-bait replicas such as worms, grubs, crayfish, minnows, and many others. Often, these replicas are enhanced, offering fluorescent colors, diverse scents, and an assortment of movements, each designed for different species and forms of fishing. The production of rubber worms is a clear example of "substitution," whereby biological entities and processes are substituted by industrial processes and/or synthetic material.[8] Producing a rubber worm with an identical use value to the nightcrawler completely eliminates the need for dairy farmers, immigrant laborers, clayey loam soil, and optimal temperatures and moisture. The largest producer of artificial baits is Berkley, which tells its customers it spares no expense on research and development to produce effective baits: "At Berkley, we know what makes a fish strike. . . . Our bait is the result of extensive research in R&D laboratories that measure fish reaction to color, scent, sight, and sound. We collect real-world data from the pros, test, test, and test again, analyze the results, and repeat."[9]

Berkley's soft baits, in particular, release water-soluble attractants that target the fish's incredibly sensitive olfactory glands. As the fake worm drifts through the water, it creates a "field of scent" that allows the fish to track the bait from a distance compared to live bait. "After thousands of tests in the lab and in the water, the right combination of components was developed to be the most appealing to the target species of fish," and the company claims their bait caught more fish than live bait "head-to-head." There are no peer-reviewed studies that conclusively back these recent marketing claims, but this has not stopped Berkley from producing an increasing array of nightcrawler substitutes, each with slightly different colors, sizes, and scents. They have even produced a "pinched nightcrawler" to mimic how anglers may "pinch" or cut up a nightcrawler into several pieces when fishing for smaller species.

Obviously, artificial baits face no restrictions and are often seen as more environmentally friendly, not only because they will not destroy forest ecosystems but also because, in the past, fish have tended *not* to swallow these artificial baits whole, thereby increasing the survival rate of the fish for anglers to practice "catch and release."[10] In a decades-long study in Wisconsin that compared the effectiveness of live versus artificial bait, researchers found that "a live bait restriction may be a viable tool for reducing exploitation in open-access walleye fisheries during population rehabilitation efforts while maintaining angling opportunities."[11] As good as artificial baits might be in attracting fish, nightcrawlers remain the more effective fish killers. As described in chapter 1, for those anglers looking for a meal or a beginner angler wanting immediate success, this is their benefit. For the conservation-minded catch-and-release anglers up for the challenge of catching a fish on a barbless hook with artificial bait, dew worms are part of the problem, which a synthetic substitute bait can solve.

SCARCE PICKING NIGHTS AND FISHING DAYS

Complicating all these challenges is climate change. As I mentioned, there are no climate skeptics among the coolers, each of whom noted how hotter and drier summers are reducing the number of optimal picking nights throughout the year. Recent studies suggest this is true. Particularly, the number of "hot nights"—defined as nighttime temperatures above 22°C—is increasing in Canada's southern regions, like southwestern Ontario.[12] As DCA Baits noted: "The worms are there. But it gets way too hot.... More and more we're really hampered by the changing environment.... We are really noticing it in our industry, how much the weather has changed in the last thirty years." It isn't just the temperature but the erratic weather patterns as well. "Big dumps of rain at one time, then nothing for three weeks? Those things kill us," DCA Baits said. This is especially frustrating for coolers who possess the necessary land and labor but cannot pick worms because of the extreme weather. As one large cooler noticed, "It would always go up and down, but it used to settle on an average." But now, in "the last year or two, the numbers [of worms picked] are dwindling."

Climate change also affects nightcrawler demand. As one cooler noted, "When it's that hot out, no one wants to be out on a boat fishing." More systematic studies note how "extreme heat significantly reduces outdoor recreational activities," like recreational angling.[13] Nor would I recommend fishing during California wildfires, Louisiana floods, or Midwestern tornados, all of which are projected to increase in the coming decades. Even in places like northern Ontario, where temperatures have not sufficiently deviated sufficiently from historical norms to a degree that would disrupt recreational fishing, climate change will likely shorten the fishing season for popular species such as the worm-hungry walleye and lake trout.[14] Indeed, the transformation of fish habitats and species distribution is one of the more direct impacts of climate change on recreational fisheries.[15] One model suggests changing temperatures and precipitation will seriously impact stream temperatures, flows, and thermal habitats across the United States,[16] putting the future of recreational fishing in peril; by the year 2100, cold-water fisheries would decline by 50 percent and reduce the aggregate number of fishing days by 6.42 million.[17] If kids today are not already taking nightcrawlers out to nearby lakes, then their future grandkids and great-grandkids certainly won't be doing so when recreational angling may be largely confined to the mountainous areas in the western United States, small pockets of New England, and the Appalachian mountains.

· · ·

COVID-19 created a sudden increase in recreational fishing and nightcrawler demand. These new and returning anglers fishing with worms have propped up the nightcrawler industry over the past two years, postponing the impact of scarce labor, land, demand, and climate change. Whether these new and returning anglers continue to fish and pay high prices for nightcrawlers, however, is an open question. The nightcrawler capitalists' uneven ability to sufficiently control the conditions and elements perpetually threatens future accumulation. Using the bait industry as an example, we can see how different industries rely on a confluence of evolving factors over which capital possesses varying degrees of control. Especially for industries that have difficulties in seizing control over the forces and conditions of production, surplus value production rests on the contextual alignment of human and nonhuman elements, each with their own tendencies, agencies, particularities, and potentialities. The nightcrawler assemblage has proved to be "durable"[18] over the course of eighty years, swapping soils, landscapes, equipment, and people in and out of the assemblage to maintain and increase surplus value. But it has not been stable or fixed. The constitutive elements of the assemblage—the (mostly) immigrant workers, the policies, the farmers, the soil, the worms, the anglers, the climate—are currently pulling in different directions. There has been no new wave of cheap labor or cheap land. A changing climate is restricting optimal picking nights (and fishing days) throughout the year. The only "positive" surprise has been the increased demand for nightcrawlers during the pandemic. Had I written this in 2019, my prediction would have been that the market would continue to shrink along with the number of coolers and pickers. COVID-19 provided an unexpected but welcome surprise for the nightcrawler industry as people sought out recreational activities of decades past that had the ability to provide social distance. The increase in recreational anglers has—for the moment—provided a boost to an increasingly fragile assemblage. Whether these new anglers continue taking trips to rivers and lakes and thereby help to stabilize the industry in the near future is the question on each cooler's mind.

"Closing the File on This Subject Forever"

As far as I can judge, it will be a curious little book. The subject has been to me a hobby-horse, and I have perhaps treated it in foolish detail.

CHARLES DARWIN on nightcrawlers

BEN WAS WEARING A SUIT IN his office, sitting behind a large oak desk. He had made a lot of money in the worm business and appeared to be a very busy man. He received several phone calls throughout our interview, which he would answer in his thick accent and animated voice. "Everyone wants to talk to me," he said as he hung up the phone. The interview was a spatial-temporal hodgepodge of worm-related stories, which he was eager to share. He told me about the time he argued with a PETA member suggesting worms should be ethically packaged five to the box. His response? "Lady, I'll put one in a box if you send me the money." He told me about the time his worms were used as an ingredient in pâté at the famous Moulin Rouge restaurant in Paris. I couldn't verify this story, but there is *some* sort of story, considering the labels on most boxes of nightcrawlers provide the warning *NOT FOR HUMAN CONSUMPTION*. He clearly enjoyed being a character in the history of the worm industry and had no plans to retire. "They found out if you retire, your brain shuts down. You gotta keep busy. My father used to say that too. I said, 'What the hell you talking about?' And he lived to be ninety-seven." He may not have the energy he once did, but that doesn't stop him. "I'm like a water buffalo, slowly pulling the water." Nico—the cooler who often fought with Ben at the early morning worm auctions—also has no plans to retire. "What the hell, I'm seventy-one. [Am I going to] give up because I can't make it anymore? ... What the hell I'm going to do tomorrow? I'm going to sit and watch the ceiling." When he does retire, he is certain he will find his customers reputable new suppliers: "I'm not going to let them without worms."

There is a definite pride when these coolers speak about the hundreds of millions of worms that have gone through their warehouses, along with the number of houses and kids' educations financed by the pennies embedded in each worm. They also enjoy telling stories about how they outmaneuvered their competitors, their first time picking worms, and their immigrant experience in Canada. There was a nostalgia for the belle epoque of worm picking in the 1970s and 1980s. As one cooler told me, "It was fun back then . . . but now it's different." The ensemble cast of human and nonhuman characters are being squeezed and pulled in different directions, making it harder and harder to get the billions of fat nightcrawlers to market. At the outset of this book, I responded to Apollo Baits that I wouldn't set foot into the business. Why would anyone? Rents continue to climb because farmers have nothing to lose. The pickers are getting older with each passing year. High nighttime temperatures in the summer limit picking days, reducing the number of worms for anglers, wages for pickers, and profit for coolers. Add to all this the tax evasion, bags of money, and cash payments that are necessary to eke out a slight advantage in a fiercely competitive industry, and you can see why I would not step foot in the industry. I would advise you to avoid it as well.

I place most of the blame for these challenges facing the industry on the protagonist of the story, *L. terrestris*. It doesn't behave passively as a good commodity should. These earthworms are subsumed in the process of capital accumulation, but only in the formal sense, not in the real, capitalist-proper sense. The conceptual interplay between the formal and real subsumption of labor and nature helps us understand how and why certain transformations of nature produce alternative, new, or peculiar labor configurations and, conversely, how the inability to radically transform or manipulate ecological processes to increase surplus value constrains such configurations. No individual set out to make Toronto the worm capital of the world; it emerged through the competitive pursuit of surplus value set in a context of socio-ecological conditions that could not easily be manipulated and rendered subject to the control of nightcrawler capitalists.

The nightcrawler trade is, in many ways, a marginal industry. It appears as an exception to the rule of innovative capitalist development. No one seems to notice it, let alone research it. Even coolers who make their living through the trade questioned why I would devote so much time to studying their business. I told DCA Baits that I've been curious about worm pickers from the first time I saw them in the middle of the night over twenty years ago. When I found no in-depth study of the industry or the actors, I told him that such a multimillion-dollar industry deserves to be uncovered, examined, and explained. I thought we might learn something from a marginal case study about how money is made and nature is transformed. Perhaps worm picking is simply an aberration, an outlying oddity, or an exception to capitalist development. Or perhaps understanding such outliers helps clarify and sharpen our conceptual tools. Uncovering what is hidden and unknown allows us to see what is theoretically useful, what is not, and what needs refining. I suggested to the cooler that I might even write a book about it. He laughed. "A book about worm picking! Good luck with that." And here we are.

The nightcrawler industry is but one case study that documents how the formal and real subsumption of both labor and nature are dialectically related, co-constituting, and maintained by the extent to which the collective agency called "capital" can assemble heterogeneous human and nonhuman actors to produce surplus value. The peculiar characteristics of nightcrawler production—why it is an underground economy, why it harvests worms from dairy farmers' fields, why labor is composed of first-generation Southeast Asian immigrants paid by the piece—can be explained by the relative inability of nightcrawler capitalists to take hold of and transform the socio-ecological conditions of production. In response, the industry has organized a disparate assemblage of human and nonhuman actors and arranged them, sometimes accidentally or unintentionally, in surprising ways to sustain the capture of surplus value over eight decades.

Why does this worm story matter? I am reminded of an E. P. Thompson line in the conclusion of *Whigs and Hunters*, where he questions the effort he put into his research and writing—"five years of notes, xeroxes, rejected drafts"—to understand what appeared to be a trivial, niche set of laws in late-1700s England. His response (and mine): "I am disposed to think that it does matter; I have a vested interest (in five years of labor) to think it may."[1] Deploying a subsumption framework to study the seemingly insignificant, almost risible, nightcrawler industry putzing around on farmer fields in one small section of the globe sheds some surprising light on the broader relationship between capitalism and nature. At its core, this book tries to explain and understand the peculiarities of the nightcrawler industry, while suggesting that the theoretical insights might be applied more broadly. Seeing how the subsumption of nature relates to the subsumption of labor can help explain why certain industries operate the way they do, why and how investments are made, why and how land is rented, and, critically, how and why labor is deployed the way it is, and who does the laboring.

In one way, the recalcitrant nightcrawler commodity appears as a prime example of the utility of ANT, assemblage, posthumanist, and critical animal frameworks that reveal the constitutive nature of nature, where the nonhuman entities and processes possess an ontology akin to their human counterparts. Indeed, most of the elements of the nightcrawler assemblage largely exist and operate autonomously from the industry. The nightcrawler capitalists didn't put the worms in the ground, nor do they shape soil structure or climatic conditions. Dairy farmers are just dairy farmers and do not need to engage with the industry. Even the bait commodity can be substituted by synthetic worms or a more controllable epigeic bait. The nightcrawler commodity only emerges and hangs together in the interstices of these other industries and processes. It is merely one element among many. To crudely oversimplify my analysis, the nightcrawler capitalists are not the boss; the worm is the boss. Well, perhaps boss is the wrong word—but it certainly exerts a hell of a lot of influence over the production process for such a seemingly inconsequential organism.

However, leveling the ontological playing field risks substituting thick description for analysis and explanation. The assemblage concept is helpful in redirecting attention to the marginalized elements of a specific phenomenon, but it doesn't always explain how or why such elements are stitched together or what is driving the assemblage forward. There is a reason why Oregonian worm pickers in the 1970s lost their jobs to immigrants in Toronto. There is a reason why nightcrawlers are picked from southwestern Ontario dairy farms and not Wisconsin dairy farms. There is a reason why worms are picked by a racialized immigrant group and not a bunch of college kids looking for a summer job. The reason is the systemic need to accumulate surplus value. Capital is not deterministic, or teleological, but experimental. It adapts, enrolls, experiments, and transforms the elements capable of increasing productivity to pump out more surplus value.

A conjoined subsumption framework ties together the assemblage concept with a focus on capital accumulation to "track the heterogeneous and plural entities *that produce surplus*, participate in and constitute the economic," which are necessary "if one is better to understand how life itself becomes a locus of accumulation" (emphasis mine).[2] Capital should not be reduced to merely one element among many; understood as "value in motion," capital accumulation is the glue that renders the assemblage coherent even while it simultaneously struggles to take hold of the socio-ecological conditions of production. Capitalists cannot dictate the conditions and elements of production. They must subsume and experiment with the people, places, and things available to them. This creates heterogeneous labor arrangements, relations, and forms of production. Through a conjoined subsumption framework, we can track the nonteleological myriad of ways capital accumulation attempts, fails, or succeeds in co-constituting labor and nature in ways necessary to increase productivity. Surplus value rests on productivity, and productivity rests on the historical ways capital exerts control and design over human and nonhuman nature. When capital fails to accumulate, the nightcrawler assemblage will fall apart.

The nightcrawler story exemplifies how reactive and experimental capital is. It firmly embeds the diversity of environments and forms of labor within capitalist relations even if capital cannot fully control and design the production process in ways it would prefer. A conjoined subsumption framework helps identify, connect, and analyze the socio-ecological conditions that co-shape the "indeterminacy of labor"[3] with the "indeterminacy of ecology."[4] Capitalist strategies to manage, coordinate, coerce, and discipline labor are never readily available or predetermined but are shaped by "lively," "unruly," or "uncooperative" biophysical characteristics of production.[5]

This has significant ramifications for understanding worker agency. While resource geography has shown how capital investments and labor must be oriented around the materiality of production, there has been less work on how varying control and design of ecological conditions of production shape forms of worker agency, meanings, and identities within the labor process. I suggest a conjoined subsumption of labor and nature framework helps explain the relationship between the socio-ecological conditions of production and workers' "constrained agency" in the worm-picking industry. Pickers retain significant control over the labor process and possess surprising agencies because the nightcrawler capitalists have been unable to radically transform the ecological conditions of nightcrawler production. Stuck in this formal subsumption of nature, the nightcrawler capitalists are also stuck with formally subsumed laborers with an idiosyncratic or "traditional" skill set neither designed nor fully controlled by the capitalist. This relationality between the control and design over nature and the control and design over labor plays a fundamental role in setting the conditions of work and shaping worker agency in the labor process. Worker agency is tied to the ecology of production.

The dialectical relationship between the subsumption of labor and nature also helps evaporate the categories of "human" and "nature"—foregrounding how the metabolic relation between humans and nature is constitutive of both. The nightcrawler is not inherently a commodity, nor is the farmer inherently a rentier. Nor, for that matter, is there such ahistorical employment that we might call a worm picker—these are created and maintained through the human/nature metabolism under capitalism.

The nightcrawler story shows how capitalism relies on noncommodified biophysical processes and elements to produce surplus value. It forces researchers to look at the material entanglements that are constitutive of the production of surplus value, and its impacts on production processes and the deployment of labor; how capital confronts nature is consequential to how value is produced and appropriated. Capital feeds off of noncommodified life forces to sustain itself, whether this is a complex human being (valued or not, waged or unwaged) or a simple earthworm. By focusing on how surplus value is created, maintained, or increased, a conjoined subsumption framework also provides an alternative opening to debates over value theory. Do plants or animals produce capitalist value alongside their human counterparts? What about photosynthesis, tectonic plates, wind speed, or UV light? I have shown how capitalist value cannot exist without subsuming nature through various forms and logics of surplus value. The extent to which capitalists harness and transform these biophysical processes (in which category I also include human beings) will affect productivity and hence rates of surplus value. Viewing the control and (re)design of labor and nature as ontologically intertwined processes hints that the standard measure of value—abstract socially necessary labor time—might be better understood as *abstract socio-ecologically necessary labor time*, in that *labor time* remains the primary metric of capitalist value but is dependent on the *socio-ecological* metabolic human/nature relation.

There is also much room for further theoretical development. The night-crawler assemblage is an example of an industry stuck operating in the formal subsumption of nature and labor, where surplus value is extracted through provisionally aligned heterogeneous human and nonhuman actors. While the conceptual interplay was useful in explaining the peculiarities of the nightcrawler commodities, the conjoined subsumption framework can also be applied to other industries that might occupy different positions along the subsumption spectrum. For example, how will the production of synthetic diamonds affect the labor regime of those working in diamond mines (and affect diamond giants like De Beers)? How might the development of perennial wheat seeds disrupt agribusiness control over agriculture? How might mushroom-picking machines affect minimum-wage employment—or perhaps, how might foreign temporary-worker programs stunt such expensive technological innovation? How can lab-grown meat affect labor and land regimes associated with meat production and climate change? Identifying the extent to which capitalists can control the ecological conditions of production in these industries provides a clearer understanding of the production of surplus value, rent relations, and capital investments, as well as the composition and constitution of labor.

Often, the commodification of nature literature describes such biophysical commodities as unruly, uncooperative, or recalcitrant to the capitalist imperatives of increasing productivity, reducing turnover time, and eliminating frictions that arise from production, exchange, and consumption. Industries that confront biophysical conditions, inputs, and processes directly in production are not special in some ontological way, but they are, as Karl Kautsky observes, "especially complex and diverse," especially for a theoretician to explain.[6] And yet, in Ontario, worms are picked, stocked, shipped, bought, and impaled on a hook with significant profit made along the way. What constrains capitalist objectives is not simply that these are unruly commodities with their own biological, physiological, and ecological characteristics; the problem, from the standpoint of capitalists, is rather that nightcrawler production is stuck in archaic production forms producing profit through the logic of absolute surplus value. Worms *become unruly* when logics of absolute surplus value appropriation hit certain barriers and the material characteristics of the commodity prevent a transition to the real subsumption of nature. In this sense, nightcrawlers were not "uncooperative" when there was growing market demand, minimal land rents, and a large immigrant labor pool. Only when these elements transformed and started impacting surplus value did we see the recalcitrance of nightcrawlers to the imperatives of capital accumulation.

. . .

In the nightcrawler industry, the "hidden abode of production" is quite literally hidden and largely unknown, and perhaps the most practical implication of this study is to bring these nightcrawlers to the surface and provide a better understanding of the economic, ecological, and social elements that constitute this multimillion-dollar industry in southwestern Ontario. DCA Baits was one cooler that welcomed increased visibility. He was frustrated with the lack of institutional support for his business, especially compared to various fruit growers. In the past, he had tried to engage in government ministries and agricultural universities: "I remember going in there, it was almost like a joke. Like, 'you guys are the worm people. . . . *Haha.*' Honestly, I was very disappointed." But the value of worms is no joke. Based on export records and domestic demand, the nightcrawlers are probably worth over $60 million to the region. DCA Baits had reason to be disappointed, as his industry was on par with other Ontario farm-based industries such as peaches, strawberries, or blueberries, all of which can access numerous avenues of state support.

Worms also appear to be the most valuable cash crop grown in the province, generating more money than farmers thought possible. Green beans are another valuable crop, and though they may resemble the green Nitro Worms, they are not nearly as valuable. The rental rates were surprising to many of the farmers I spoke with who had stayed clear of or hadn't engaged with the industry. This book offers a bit more clarity to farmers, government officials, recreational anglers, and anyone else who passes a funny-looking transport truck on Highway 401 and wonders about the different actors in the industry, their roles, and the myriad of challenges they face.

L. terrestris is one of the most researched worms in the world. But if you glance at the soil science journal articles, you will quickly notice most research is conducted on a certain kind of *L. terrestris*. "*Lumbricus terrestris* were purchased from a local bait store" or "*Lumbricus terrestris* were commercially purchased" are common phrases in the methodology sections of the articles. These are not just any dew worms; researchers are studying the Canadian nightcrawler commodity. As such, this nightcrawler story offers practical implications and life histories for their research specimens.[7] Christopher Lowe and Kevin Butt had previously questioned the use of such earthworms for research and practical applications because information about their place of origin and environmental exposures was unknown. They suggest culturing *L. terrestris* might be more efficient than procuring the worm through the bait market, which they assume to be "less sustainable and not as effective as earthworm culture."[8] This book offers scientific researchers too a better understanding of the worms they use, the processes of nightcrawler production, and the environmental settings and life histories of the worm. In particular, the fecundity of nightcrawlers in dairy farms appears to be a natural experiment in biostimulation, that is, finding ways to manipulate the on-site biophysical elements to stimulate earthworm reproduction. And—at least in the near future—procuring *L. terrestris* through the nightcrawler industry is still more economical and efficient than culturing the worms to serve various research or environmental remediation demands. It also appears to be surprisingly sustainable, considering the conditions on the farm that enable pickers to harvest worms from the same field each year. At the same time, these Ontario *L. terrestris* have a unique life history through their incorporation into Ontario dairy farming. The Canadian nightcrawlers, I would suggest, are a tad spoiled, growing exceptionally large, living in luxury in no-till alfalfa fields, and consuming a smorgasbord of manure feedstock. It may be problematic to assume these Canadian nightcrawlers are representative of the average, or typical, *L. terrestris*. In a recent study, Dutch researchers inoculated pastures in the Netherlands with store-bought Canadian nightcrawlers. The survival rate after eighteen months was an underwhelming 6 percent: "*L. terrestris* mortality may have been related to their origin and life history."[9] I would suspect the Canadian nightcrawlers may not be the heartiest specimens for certain applications.

Perhaps the most surprising observation was the impact of removing worms from the field of farmers engaged in full-year contracts. Worm picking does not appear to affect yields, and counterintuitively could potentially improve them. This deserves more research both in controlled environments and in field experimentation, as it has the potential to allow farmers to leave their fields fallow while profiting at the same time. This could easily incentivize more farmers to engage in the industry, even those outside of the dairy industry. No-till cash-croppers, for example, could pay for manure and still handsomely profit by giving their land a rest for a year while pickers uproot the largest *L. terrestris*. Future earthworm research can also examine how the massive removal of the largest adult *L. terrestris* worms affects trophic relations in an agricultural setting, a phenomenon that has not been studied. How does such widespread removal kick-start and speed up the growth and reproduction of juveniles? How does removing the largest macrofauna reorient other earthworm species and other soil biota? Likewise, as many farmers questioned, what is the nutrient value of the earthworm itself? If corn takes out so much nitrogen from the soil, what nutrients or chemicals are removed with 200,000 worms per acre? The rapid consumption of manure by *L. terrestris* is well known, but more research could be done on the speed at which they digest different types of manures and make nutrients available through their castings. This could be particularly useful for jurisdictions struggling to manage large quantities of manures relative to their land base.

In the social domain, worm picking is a unique job whose surface I have only skimmed. The industry's underground nature, combined with my positionality as a white, male, English-speaking Canadian researcher, made it difficult for me to expand on the experience of worm pickers—something that I believe merits a more grounded ethnographic study. Historically, worm picking has been tied to the immigrant experience of Toronto and so few of these stories have been told.[10] The physicality of the job, the aging immigrant workforce, and the relatively high pay combined with the surprising expressions of freedom signal there is much more research that could inform discussions about how recent immigrants adapt to Canadian cities, how they navigate employment opportunities, and how the nature of the production process is constitutive of their work and the immigrant experience.

Indeed, the ecological underpinnings of production have largely been ignored for the role they play in politicizing the labor process and shaping worker strategies. What has only been implicit in my analysis is politics. This is not a book that will conclude with the statement "What is to be done," nor is it a muckraking account of exploitative labor conditions or soil degradation (which, quite honestly, I thought it might be). I am not calling for the abolition of the industry, regulations to restrict earthworm removal, or a consumer boycott of the Canadian nightcrawler. I doubt "fair trade" worms would catch on, nor would I consider any of these worms "organic." Such calls are too simple and do not take into account the dialectical dynamics at play. Instead, my analysis has shown how each actor in the industry can exercise power differentials that socio-ecological conditions of production have afforded them. Actions taken around one element of the assemblage will reverberate through the whole. Analyzing this peculiar assemblage, taking account of nonhuman elements often left on the analytical cutting-room floor, and piecing it back together reveals alternative routes for human (and nonhuman) struggles.

On the one hand, what should be clear is that the worm pickers get the least by physically doing the most, while dairy farmers get the most by doing the least. At the outset, it appears to be a formulaic class antagonism between labor, capital, and rentiers—typical relations that we would assume could be addressed through collective organizing. And yet, at the current conjuncture, worm pickers are not afraid to use their positions to negotiate better wages and jump between coolers that offer the highest piece rates. They can also retain more money by working in cash and not paying taxes. Others are working illegally in a production process that affords them considerable anonymity—quite literally working in the dark in the middle of nowhere. Such pickers have little interest in forming a union or formalizing the unregulated industry, as it would restrict certain "freedoms" they believe worm picking has afforded them.

If we want to improve the wages and conditions of pickers, we have to consider the relations between the elements in the assemblage—between production, distribution, and consumption. If people are willing to pay $6 US per dozen worms, as they did during the 2020 fishing season, wages can (and did) rise. When I first conducted preliminary interviews the average piece rate was $25. Post-lockdown, it has stabilized around $35. At the same time, if more farmers could be convinced that devoting land to worm picking could provide ecological benefits to the soil, rents could diminish, and the savings, I suspect, would not stuff the pockets of the nightcrawler capitalists but would likely increase picker wages. Such an action would not come from the goodness of their own hearts but out of the necessity to attract more productive labor away from their competitors. Improving pickers' conditions could also come through broader state regulation as opposed to workplace organizing. Advocates for temporary migrants have steadfastly called for changes to the structural and racial barriers to residency status—something that continues to be hotly debated in Canada.[11] Particularly for migrant workers, this would increase worker mobility, options for formal payment, and access to social welfare benefits, all while maintaining forms of "freedom" they express in the labor process. Viewing the formal and real subsumption of labor and nature as a conjoined process reveals how the extent to which capital can control and design ecological conditions of production will shape their strategies of control and design over the labor process and shape worker expressions of freedom and agency in the workplace. In this sense, we can see how the fight over surplus is a simultaneously political and ecological struggle that has developed through the entanglement of different elements into the nightcrawler assemblage over the past eight decades.

· · ·

In 1971, journalist John Miller found himself reporting on the hidden nightcrawler industry. He spoke with many of the early Torontonian worm barons, touching on earthworm biology, humidity, labor relations, and export markets. He concluded his long and informative newspaper article as follows: "There is no large body of literature available on worms, or for those who are (or would be) worm pickers. This short piece, then, I hope, may prompt some biologist to embark upon a more sustained flight, leaving us with a work that even the slowest of minds can understand, and, hopefully, closing the file on this subject forever."[12]

Well, I'm not a biologist, but I have done my best to "embark upon a more sustained flight" regarding this peculiar industry than others have previously attempted. The simple act of getting nightcrawlers on an angler's hook requires a contingent arrangement of various historical, social, and political factors. And despite the lack of clear, powerful agents or actants in the nightcrawler industry, there still remains the internal logic of capital accumulation that manages to seize the limited opportunities that arise in order to increase labor productivity and continue to generate surplus value. Marx's oft-cited dictum of capital's drive for "accumulation for the sake of accumulation, production for the sake of production," should never leave out what exactly is being accumulated or produced: surplus value.

My study shows how this underground economy represents a rather ad hoc, accidental, interstitial, contingent, hodgepodge capitalism—but a capitalism nonetheless. Challenges to capital accumulation in the industry are overcome not by entrepreneurial creativity, human initiatives, or technological innovation but through a shifting assemblage capable of cobbling together the necessary soil, bodies, markets, and biophysical processes. Despite its inner logic to accumulate, capital is not omnipotent. Assemblages are contingent, fragile, and dependent on how the collective agency of capital simultaneously subsumes labor and nature but in limited and uneven ways. As we can see in the nightcrawler industry, minor disruptions in one element can crack the coherence of the assemblage as a whole. If the nightcrawler assemblage can't be reknit, capital accumulation slows. If younger anglers can't be found, if new waves of immigrants don't take to picking, if climate change reduces the number of productive picking nights, if bait regulations restrict the movement of worms, if dairy farmers don't rent their land, then all the autonomous elements that have hitherto maintained the nightcrawler assemblage may go their separate ways. This would be disastrous for the industry but perhaps beneficial for the protagonist. I would prefer not to speak for the worms, but I'm confident they would rather enjoy living out their lives comfortably nestled in their no-till protected burrows, feasting on cow manure, and mating with their friendly neighbors than ending up on a hook. I suspect they would happily shed their "Canadian nightcrawler" title and revert back to being the lowly and humble dew worm.

NOTES

PROLOGUE

1. Cherry, "After a Fashion: Night Creature Not Men from Mars—Just Worm Pickers."

2. *Globe and Mail,* "Human Robins: Fast Worm-Picker Can Make Good Money, Live Bait Executive Says."

3. Webster, "Willing Worms Wait for Warm, Windless Weather."

4. Egan, "Crawler Shock Hitting Local Anglers Right in the Worm Bucket."

5. Taekema, "Worm Picking a Slippery Industry in Toronto."

6. Taekema, "Worm Picking a Slippery Industry in Toronto."

7. Throughout the book, dollar figures are assumed to be in Canadian dollars. I will specify figures in USD when appropriate, as I do below, as the majority of night-crawlers are purchased in the United States.

8. Decision No. 5881/89, Ontario Workplace Safety and Insurance Appeals Tribunal, 1991.

9. My methods for calculating these numbers are rather *grosso modo.* Most coolers suggest between 500 and 700 million, based on their own sales and estimates of their competitors. This includes a few coolers that might sell between 80 and 100 million worms annually, along with several medium-sized coolers selling 20 to 60 million and smaller coolers with under 10 million. I estimate the wholesale value— the amount of money that stays in the region—based on HS Code 010690 "Live Bait." Americans are the primary purchasers of the worms, at $60 million, with Europeans purchasing around $7 million (Government of Canada, "Tillage and Seeding Practices, Census of Agriculture, 2021"). Not included in these numbers are the worms sold in Canada, which coolers suggest is about 10 percent of the market, thus increasing the wholesale value by $4 to $5 million. This would provide a total wholesale value around $72 million.

As to the number of coolers, I could independently identify twenty-three in operation. In interviews, coolers suggested there would be another dozen or so smaller outfits, which would be extremely difficult to find. Crew chief numbers are even more difficult to find, as they prefer to avoid scrutiny for reasons described in chapter 4. Based on the total volume of worms picked and interviews with coolers who stated the number of crew chiefs they use, there are likely between twenty-five and thirty-five independent drivers.

We can estimate the number of pickers and their annual remuneration based on the total number of worms picked, an average piece rate of $35 per thousand worms, and interviews with coolers who suggest an average picker on an average picking night would collect 6,400 worms, or 640,000 worms picking 100 nights. Using these numbers, I estimate there are between 780 and 1,100 pickers making an average of $22,000 per year.

10. There are several earthworm species that have "nightcrawler" in the common name. When I use the term nightcrawler without further specification, I am referring to *L. terrestris*.

11. Keller et al., "From Bait Shops to the Forest Floor: Earthworm Use and Disposal by Anglers."

12. Champ, "Nightcrawlers Make Good Bait in Warm Water."

INTRODUCTION

1. O. Jackson, "Worm Trap."

2. O. Jackson, "Worm Trap."

3. Kautsky, *The Agrarian Question: In Two Volumes*.

4. Bernstein, "Is There an Agrarian Question in the 21st Century?"; Akram-Lodhi and Kay, "Surveying the Agrarian Question (Part 2): Current Debates and Beyond."

5. Moore, "Ecological Crises and the Agrarian Question in World-Historical Perspective"; Akram-Lodhi and Kay, "Surveying the Agrarian Question (Part 2): Current Debates and Beyond."

6. Foster, "Marx's Theory of Metabolic Rift: Classical Foundations for Environmental Sociology"; Schneider and McMichael, "Deepening, and Repairing, the Metabolic Rift."

7. Mann and Dickinson, "Obstacles to the Development of a Capitalist Agriculture"; Goodman, Sorj, and Wilkinson, *From Farming to Biotechnology: A Theory of Agro-Industrial Development*; Boyd, Prudham, and Schurman, "Industrial Dynamics and the Problem of Nature."

8. Goodman, Sorj, and Wilkinson, *From Farming to Biotechnology: A Theory of Agro-Industrial Development*.

9. Henderson, "Nature and Fictitious Capital: The Historical Geography of an Agrarian Question"; Felli, "On Climate Rent"; Ouma, Johnson, and Bigger, "Rethinking the Financialization of 'Nature.'"

10. Kirsch, "Cultural Geography I: Materialist Turns"; Bakker and Bridge, "Material Worlds? Resource Geographies and the Matter of Nature."

11. Collard and Dempsey, "Life for Sale? The Politics of Lively Commodities"; Watts and Scales, "Seeds, Agricultural Systems and Socio-natures: Towards an Actor–Network Theory Informed Political Ecology of Agriculture."

12. White, *The Organic Machine: The Remaking of the Columbia River*.

13. Darwin, *The Formation of Vegetable Mould, Through the Action of Worms: With Observations on Their Habits*, 313.

14. Barua, "Animating Capital: Work, Commodities, Circulation."

15. Leff, *Green Production: Toward an Environmental Rationality*; McCarthy, "Political Ecology/Economy."

16. Guthman, *Wilted: Pathogens, Chemicals, and the Fragile Future of the Strawberry Industry*.

17. For a summary of these critiques and responses see Foster, Clark, and York, *The Ecological Rift: Capitalism's War on the Earth*.

18. Walker, "Value and Nature: From Value Theory to the Fate of the Earth."

19. Kallis and Swyngedouw, "Do Bees Produce Value? A Conversation between an Ecological Economist and a Marxist Geographer."

20. Moore, *Capitalism in the Web of Life: Ecology and the Accumulation of Capital*.

21. Federici, *Caliban and the Witch*; Mies, *Patriarchy and Accumulation on a World Scale: Women in the International Division of Labour*.

22. Walker, "Value and Nature: Rethinking Capitalist Exploitation and Expansion," 55.

23. Prudham, *Knock on Wood: Nature as Commodity in Douglas-Fir Country*, 8.

24. Burkett, "Value, Capital and Nature: Some Ecological Implications of Marx's Critique of Political Economy," 340.

25. Murray details two other forms of subsumption in Marx's writings, hybrid and ideal subsumption. For a detailed discussion see Murray, "The Social and Material Transformation of Production by Capital: Formal and Real Subsumption in *Capital*."

26. Marx, *Capital: Volume I*, 506–8.

27. Marx, *Capital: Volume I*, 361.

28. Obviously, the logics of absolute and relative surplus value are not mutually exclusive, and seeing them as such might trip up commentators (e.g., John Stuart Mill) who think massive increases in productivity would reduce necessary human labor—permitting increased leisure time for workers. Such a notion, however, forgets that increasing productivity *and* employing more workers to work longer hours would increase profit even further!

29. Boyd, Prudham, and Schurman, "Industrial Dynamics and the Problem of Nature."

30. Goodman, Sorj, and Wilkinson, *From Farming to Biotechnology: A Theory of Agro-Industrial Development*.

31. Mann and Dickinson, "Obstacles to the Development of a Capitalist Agriculture."

32. Prudham, *Knock on Wood: Nature as Commodity in Douglas-Fir Country*.

33. Kloppenburg, *First the Seed: The Political Economy of Plant Biotechnology*.

34. Guthman, *Wilted: Pathogens, Chemicals, and the Fragile Future of the Strawberry Industry*.

35. W. Boyd, "Making Meat: Science, Technology, and American Poultry Production"; M. Cooper, "Open up and Say 'Baa': Examining the Stomachs of Ruminant Livestock and the Real Subsumption of Nature."

36. W. Boyd, "Making Meat: Science, Technology, and American Poultry Production."

37. Boyd, Prudham, and Schurman, "Industrial Dynamics and the Problem of Nature," 564.

38. Boyd, Prudham, and Schurman, "Industrial Dynamics and the Problem of Nature," 564.

39. N. Smith, "Nature as Accumulation Strategy."

40. N. Smith, "Nature as Accumulation Strategy"; Carton and Andersson, "Where Forest Carbon Meets Its Maker: Forestry-Based Offsetting as the Subsumption of Nature."

41. N. Smith, *Uneven Development: Nature, Capital, and the Production of Space*.

42. D. Mitchell, "Dead Labor and the Political Economy of Landscape—California Living, California Dying."

43. Carton and Andersson, "Where Forest Carbon Meets Its Maker: Forestry-Based Offsetting as the Subsumption of Nature."

44. Boyd and Prudham, "On the Themed Collection: The Formal and Real Subsumption of Nature."

45. Tsing, *The Mushroom at the End of the World: On the Possibility of Life in Capitalist Ruins*.

46. Williams, "How Are Fisheries and Aquaculture Institutions Considering Gender Issues?"; Szymkowiak, "Genderizing Fisheries: Assessing over Thirty Years of Women's Participation in Alaska Fisheries."

47. Guthman, *Wilted: Pathogens, Chemicals, and the Fragile Future of the Strawberry Industry*.

48. Braun, "The 2013 Antipode RGS-IBG Lecture: New Materialisms and Neoliberal Natures."

49. T. Mitchell, *Rule of Experts: Egypt, Techno-Politics, Modernity*.

50. Bennett, *Vibrant Matter: A Political Ecology of Things*, 97.

51. Callon, "Economic Markets and the Rise of Interactive Agencements: From Prosthetic Agencies to Habilitated Agencies"; Bennett, *Vibrant Matter: A Political Ecology of Things*; Anderson and McFarlane, "Assemblage and Geography"; Müller, "Assemblages and Actor-networks: Rethinking Socio-material Power, Politics and Space"; Tsing, *The Mushroom at the End of the World: On the Possibility of Life in Capitalist Ruins*.

52. Müller, "Assemblages and Actor-Networks: Rethinking Socio-material Power, Politics and Space," 28.

53. Mitchell, *Rule of Experts: Egypt, Techno-Politics, Modernity*; Gibson-Graham, *A Postcapitalist Politics*.

54. Fine, "From Actor-Network Theory to Political Economy."

55. Castree, "False Antitheses? Marxism, Nature and Actor-networks," 123. See also Kirsch and Mitchell, "The Nature of Things: Dead Labor, Nonhuman Actors, and the Persistence of Marxism"

56. Tsing, *The Mushroom at the End of the World: On the Possibility of Life in Capitalist Ruins*, 23.

57. Guthman, *Wilted: Pathogens, Chemicals, and the Fragile Future of the Strawberry Industry*, 18.

58. Barua, "Animating Capital: Work, Commodities, Circulation."

59. MacDonald, *Aspects of Iu-Mien Refugee Identity*; Pfeifer, "Community, Adaptation, and the Vietnamese in Toronto"; Preibisch, "Pick-Your-Own Labor: Migrant Workers and Flexibility in Canadian Agriculture 1."

60. Tomlin, "The Earthworm Bait Market in North America."

CHAPTER 1

Portions of this chapter were previously published in Steckley, "Nightcrawler Commodities: A Brief History on the Commodification of the Humble Dew Worm."

1. Lamphier, "What Is Our Man Doing on Golf Course at 4 in the Morning?," 11.

2. In the *Grundrisse*, however, he specifically addresses the historicity of consumption and the social meaning of the commodity. "Production not only supplies a material for the need, but it also supplies a need for the material . . . The need which consumption feels for the object is created by the perception of it" (Marx, *Capital: Volume I*, 92).

3. Cook et al., *Cultural Turns/Geographical Turns: Perspectives on Cultural Geography*.

4. Gregson, "And Now It's All Consumption?"; Goss, "Geography of Consumption I."

5. Wood, *The Origin of Capitalism: A Longer View*, 75.

6. Marx, *Capital: Volume I*, 433.

7. Sahrhage and Lundbeck, *A History of Fishing*.

8. Pitcher and Hollingworth, "Fishing for Fun: Where's the Catch?," 4.

9. Hoffmann, "Trout and Fly, Work and Play in Medieval Europe," 62.

10. Hoffmann, "Fishing for Sport in Medieval Europe: New Evidence."

11. Hoffmann, "Fishing for Sport in Medieval Europe: New Evidence," 900.

12. Berners, *An Older Form of The Treatyse of Fysshynge Wyth an Angle*, 1,37.

13. Hoffmann, "Economic Development and Aquatic Ecosystems in Medieval Europe," 654.

14. Hummel, *Hunting and Fishing for Sport: Commerce, Controversy, Popular Culture*; Herd, *The Fly*.

15. Herd, *The Fly*, 247.

16. Herd, *The Fly*; Locker, "The Social History of Coarse Angling in England AD 1750–1950."

17. Lowerson, *Sport and the English Middle Classes, 1870–1914*, 43; Herd, *The Fly*, 292.

18. Halford, *An Angler's Autobiography*, 172.

19. Locker, "The Social History of Coarse Angling in England AD 1750–1950."

20. Herd, *The Fly*.

21. *The Field*, "Worm Fishing for Fresh Trout."

22. Washabaugh and Washabaugh, *Deep Trout: Angling in Popular Culture*, 56.

23. *New York Times*, "Bait Fishing Popular," 2.

24. McLean, *A River Runs Through It and Other Stories*.

25. *Duluth New Tribune*, "The Humble Angleworm."

26. De Grazia, "Of Time, Work, and Leisure."

27. Fish and Wildlife Services, "1960 National Survey of Fishing and Hunting."

28. Fish and Wildlife Services, "1985 National Survey of Fishing and Hunting."

29. James, "When Nothing Else Works . . . Try a Worm."

30. *New York Times*, "President Bait Wins Jersey Fishing Bout."

31. Camp, "Albert, Fly Purist Take 4 Brown Trout, But His Heart Is Heavy, He Used Worms."

32. Bryant, "A Case for Fishing with Worms," 11.

33. Raines, "Fishing with Presidents."

34. Buxton, "Worm Fishing for Trout," 278.

35. James, "When Nothing Else Works . . . Try a Worm."

36. Fegely, "Canada's Drought Cuts Flow of Nightcrawlers."

37. Wolfthal, "Shedding Some Light on the Night Crawler, Still a Fine Bait."

38. Tomlin, "The Earthworm Bait Market in North America," 335.

39. James, "When Nothing Else Works . . . Try a Worm"; Bryant, "Anglers Still Favor Earthworms."

40. Crosson, "If You like Fishing, Prepare to Pay a Little More for Your Worms This Year"; Fegely, "Canada's Drought Cuts Flow of Nightcrawlers."

41. *Star Tribune*, "Business Goes as the Worms Turn."

42. Lamphier, "What Is Our Man Doing on Golf Course at 4 in the Morning?," 11.

43. *The Times-Picayune*, "Recent Discovery Ends Bait Digging. Lucedale Man Finds Method of Coaxing Angle Worms out of the Ground."

44. *New Haven Evening Register*, "Bring Out the Worms"; *Trenton Evening Times*, "Why Dig Fish Worms."

45. *The Times-Picayune*, "'Bait' Sales Make Monroe Lads Rich."

46. *New Haven Evening Register*, "Bring Out the Worms."

47. *Plain Dealer*, "Gathering Worms for Bait. Some Fishermen Have Novel Methods of Procuring A Supply."

48. *Aberdeen Daily News*, "Goin' a-Fishin'? How to Get and Keep the Bait That Gets the Fish."

49. *Kalamazoo Gazette*, "Sale of Angleworms to Fishermen Makes New Industry in City."

50. *Grand Rapids Press*, "Make Their Living by Digging Worms."

51. *New York Times*, "Shave Pays for Fish Worms."

52. *New Haven Evening Register*, "Bring Out the Worms."

53. *Plain Dealer*, "Gathering Worms for Bait. Some Fishermen Have Novel Methods of Procuring A Supply."

54. *Kalamazoo Gazette*, "Sale of Angleworms to Fishermen Makes New Industry in City."

55. *The Times-Picayune*, "'Bait' Sales Make Monroe Lads Rich."

56. *Kalamazoo Gazette*, "Kalamazoo Kiddie Makes Spending Money Selling Worms at Depots."

57. *Kalamazoo Gazette*, "Sale of Angleworms to Fishermen Makes New Industry in City."

58. *Kalamazoo Gazette*, "Kalamazoo Kiddie Makes Spending Money Selling Worms at Depots."

59. *Grand Rapids Press*, "Make Their Living by Digging Worms."

60. *Boston Daily Globe*, "Makes Money Selling Worms."

61. *New York Times*, "Barber Invents Machine to Vend His Fish Bait."

62. Broomhall, "Can't Beat Those Worms."

63. Rupp, "Hunting and Fishing."

64. *Statesman Journal*, "Worm Hunters Ruin Flowers."

65. Easterling, "Family Turns Can of Worms into Booming Business."

66. Hodge, "Night Crawler Sellers Worm in on Fun, Profit."

67. Beltaire, "Running a Golf Course Is No Picnic."

68. Cooper, "There's a Gold Mine in Worms."

69. *News-Journal*, "Classified Ad."

70. Husar, "Fish-Bait Business Is Attractive, But U.S. Farmers Aren't Biting."

71. *Minneapolis Tribune*, "Shortage of Pickers Hurts Worm Farm."

72. After the Vietnam war, refugees in Oregon provided some labor for worm picking, which has allowed the industry to persist (at however small a scale) to the present day; see MacDonald, *Aspects of Iu-Mien Refugee Identity*, 104.

73. *Daily Spectrum*, "Worm-Pickers Find a Buried Treasure"; Hodge, "Night Crawler Sellers Worm in on Fun, Profit."

74. *Times-News*, "Breathless Worms Make Student Money."

75. *Globe and Mail*, "Worm Pickers Can Make Big Pay."

76. Canadian Soil Information Service, "Soil Landscapes of Canada."

77. FAO, "Soil Map of the World."

78. Lavkulich and Arocena, "Luvisolic Soils of Canada: Genesis, Distribution, and Classification."

79. Nuutinen and Butt, "Worms from the Cold: Lumbricid Life Stages in Boreal Clay during Frost"; Tomlin and Miller, "Impact of Ring-Billed Gull (Larus

Delawarensis Ord.) Foraging on Earthworm Populations of Southwestern Ontario Agricultural Soils."

80. Callaham and Hendrix, "Relative Abundance and Seasonal Activity of Earthworms (Lumbricidae and Megascolecidae) as Determined by Hand-Sorting and Formalin Extraction in Forest Soils on the Southern Appalachian Piedmont."

81. Lamphier, "What Is Our Man Doing on Golf Course at 4 in the Morning?"; and Tomlin (personal communication).

82. M. Boyd, "Immigration Policies and Trends: A Comparison of Canada and the United States."

83. Troper, "History of Immigration to Toronto Since the Second World War: From Toronto 'the Good' to Toronto 'the World in a City,'" 10.

84. Papp-Zubrits, "The Forgotten Generation: Canada's Hungarian Refugees of 1956."

85. Newman, "The Hungarians," 14.

86. Troper, "History of Immigration to Toronto Since the Second World War: From Toronto 'the Good' to Toronto 'the World in a City,'" 12.

87. *Washington Post*, "Golf Course Gleaners Stalk Worms, Not Ball."

88. Frebman, "Rebate on Bait: Judge Decides What's Due for Dew Worm Loss"; *Globe and Mail*, "Worm Pickers Can Make Big Pay."

89. *Globe and Mail*, "How to Earn $17.50: Get Light, Hunt Worms"; Frebman, "Rebate on Bait: Judge Decides What's Due for Dew Worm Loss."

90. *Globe and Mail*, "Worm Pickers Can Make Big Pay."

91. *Globe and Mail*, "Worm Pickers Can Make Big Pay."

92. *Globe and Mail*, untitled, April 1, 1961.

93. Cherry, "After a Fashion: Night Creature Not Men from Mars—Just Worm Pickers."

94. Cherry, "After a Fashion: Night Creature Not Men from Mars—Just Worm Pickers."

95. Cherry, "After a Fashion: Night Creature Not Men from Mars—Just Worm Pickers."

96. *Globe and Mail*, "Human Robins: Fast Worm-Picker Can Make Good Money, Live Bait Executive Says."

97. *Windsor Star*, "Worms Rejoice."

98. Webster, "Willing Worms Wait for Warm, Windless Weather."

99. *Globe and Mail*, "Ontario's Night Crawlers Outclass African Giants."

100. Abel, "Subterranean Harvest."

101. Husar, "Fish-Bait Business Is Attractive, But U.S. Farmers Aren't Biting"; *Marysville Journal-Tribune*, "Return of Walleyes Brings Worm Wars"; Cooper, "There's a Gold Mine in Worms."

102. *Minneapolis Tribune*, "Shortage of Pickers Hurts Worm Farm"; Hodge, "Night Crawler Sellers Worm in on Fun, Profit."

103. K. Jackson, "Best Way to Hook Kids Is with Worms."

CHAPTER 2

1. Graff, "Darwin on Earthworms—The Contemporary Background and What the Critics Thought."

2. Montgomery, *Dirt: The Erosion of Civilizations*, 12.

3. Engel-Di Mauro, *Ecology, Soils, and the Left: An Ecosocial Approach*, 37.

4. Bouché, "Strategies lombriciennes."

5. Dominguez, "State-of-the-Art and New Perspectives on Vermicomposting Research," 404.

6. Dominguez and Edwards, "Biology and Ecology of Earthworm Species Used for Vermicomposting."

7. Capowiez, Sammartino, and Michel, "Burrow Systems of Endogeic Earthworms: Effects of Earthworm Abundance and Consequences for Soil Water Infiltration."

8. Epigeic, endogeic, and anecic are not fixed categories, as additional subcategories can be made. Some earthworm behavior may be considered epi-anecic or endo-anecic, where they exhibit traits of several of the ecotypes. See Hoeffner et al., "Epi-Anecic Rather Than Strict-Anecic Earthworms Enhance Soil Enzymatic Activities."

9. Butt and Lowe, "Controlled Cultivation of Endogeic and Anecic Earthworms."

10. Sherman, *The Worm Farmer's Handbook: Mid- to Large-Scale Vermicomposting for Farms, Businesses, Municipalities, Schools, and Institutions*, 78.

11. Hammon, "A Worm Farm Rewrites the Start-up Rules."

12. United States Court of Appeals, N. 15-5175.

13. Kammenga et al., "Explaining Density-Dependent Regulation in Earthworm Populations Using Life-History Analysis," 93.

14. Lowe and Butt, "Culture Techniques for Soil Dwelling Earthworms: A Review."

15. Lowe and Butt, "Culture Techniques for Soil Dwelling Earthworms: A Review."

16. Reimer and Chartrand, *A Historical Profile of the James Bay Area's Mixed European-Indian or Mixed European-Inuit Community*.

17. Lindroth, *The Faunal Connections between Europe and North America*, 157.

18. Bain and King, "Asylum for Wayward Immigrants: Historic Ports and Colonial Settlements in Northeast North America," 120.

19. Borron, "Report on the Basin of Moose River and Adjacent Country Belonging to the Province of Ontario," 43.

20. Reynolds, "The Distribution of Earthworms (Annelida, Oligochaeta) in North America."

21. Baden and Beekman, "Culture and Agriculture."

22. Doolittle, "Permanent vs. Shifting Cultivation in the Eastern Woodlands of North America Prior to European Contact."

23. Doolittle, "Permanent vs. Shifting Cultivation in the Eastern Woodlands of North America Prior to European Contact"; Mt. Pleasant, "A New Paradigm for Pre-Columbian Agriculture in North America"; Mt. Pleasant and Burt, "Estimating Productivity of Traditional Iroquoian Cropping Systems from Field Experiments and Historical Literature."

24. Mt. Pleasant, "The Paradox of Plows and Productivity: An Agronomic Comparison of Cereal Grain Production under Iroquois Hoe Culture and European Plow Culture in the Seventeenth and Eighteenth Centuries."

25. Telford, "How the West Was Won: Land Transactions between the Anishinabe, the Huron and the Crown in Southwestern Ontario."

26. "Current Land Claims."

27. Simner, *How Middlesex County Was Settled with Farmers, Artisans, and Capitalists: An Account of the Canada Land Company in Promoting Emigration from the British Isles in the 1830s through the 1850s.*

28. Reynolds, *The Earthworms (Lumbricidae and Sparganophilidae) of Ontario.*

29. Reynolds, "The Distribution of Earthworms (Annelida, Oligochaeta) in North America," 149.

30. Edwards, "The Importance of Earthworms as Key Representatives of the Soil Fauna," 6.

31. Andriuzzi et al., "Anecic Earthworms (Lumbricus Terrestris) Alleviate Negative Effects of Extreme Rainfall Events on Soil and Plants in Field Mesocosms."

32. Edwards, "The Importance of Earthworms as Key Representatives of the Soil Fauna."

33. Edwards, "The Importance of Earthworms as Key Representatives of the Soil Fauna," 7.

34. Furlong et al., "Molecular and Culture-Based Analyses of Prokaryotic Communities from an Agricultural Soil and the Burrows and Casts of the Earthworm Lumbricus Rubellus."

35. Bertrand et al., "Earthworm Services for Cropping Systems: A Review."

36. Binet and Trehen, "Experimental Microcosm Study of the Role of Lumbricus Terrestris (Oligochaeta: Lumbricidae) on Nitrogen Dynamics in Cultivated Soils."

37. Van Groenigen et al., "Earthworms Increase Plant Production: A Meta-Analysis."

38. Tomlin and Gore, "Effects of Six Insecticides and a Fungicide on the Numbers and Biomass of Earthworms in Pasture."

39. Carson, *Silent Spring*, 108.

40. Bellon, "In Roundup Case, U.S. Judge Cuts $2 Billion Verdict Against Bayer to $86 Million."

41. Dittbrenner, Schmitt, and Capowiez, "Sensitivity of Eisenia Fetida in Comparison to Aporrectodea Caliginosa and Lumbricus Terrestris after Imidacloprid Exposure"; Stellin et al., "Effects of Different Concentrations of Glyphosate (Roundup 360°) on Earthworms (Octodrilus Complanatus, Lumbricus Terrestris and Aporrectodea Caliginosa) in Vineyards in the North-East of Italy"; Gaupp-

Berghausen et al., "Glyphosate-Based Herbicides Reduce the Activity and Reproduction of Earthworms and Lead to Increased Soil Nutrient Concentrations."

42. Nuutinen et al., "Glyphosate Spraying and Earthworm Lumbricus Terrestris L. Activity: Evaluating Short-Term Impact in a Glasshouse Experiment Simulating Cereal Post-Harvest."

43. Curry, "Factors Affecting the Abundance of Earthworms in Soils."

44. Edwards and Lofty, "Nitrogenous Fertilizers and Earthworm Populations in Agricultural Soils."

45. Whalen, Parmelee, and Edwards, "Population Dynamics of Earthworm Communities in Corn Agroecosystems Receiving Organic or Inorganic Fertilizer Amendments."

46. Pfiffner and Luka, "Earthworm Populations in Two Low-Input Cereal Farming Systems"; Pfiffner and Mäder, "Effects of Biodynamic, Organic and Conventional Production Systems on Earthworm Populations."

47. Irmler, "Changes in Earthworm Populations during Conversion from Conventional to Organic Farming."

48. Stewart, *The Earth Moved: On the Remarkable Achievements of Earthworms*, 139.

49. Hoefer and Hartge, "Subsoil Compaction: Cause, Impact, Detection, and Prevention."

50. Montgomery, *Growing a Revolution: Bringing Our Soil Back to Life*, 20.

51. Tomlin and Fox, "Earthworms and Agricultural Systems: Status of Knowledge and Research in Canada," 273; Fox et al., "Earthworm Population Dynamics as a Consequence of Long-Term and Recently Imposed Tillage in a Clay Loam Soil."

52. Birkás et al., "Tillage Effects on Compaction, Earthworms and Other Soil Quality Indicators in Hungary."

53. Curry, "Factors Affecting the Abundance of Earthworms in Soils," 104.

54. Bertrand et al., "Earthworm Services for Cropping Systems: A Review."

55. Edwards and Lofty, "Nitrogenous Fertilizers and Earthworm Populations in Agricultural Soils"; Butt, "Food Quality Affects Production of Lumbricus Terrestris (L.) under Controlled Environmental Conditions"; Onrust and Piersma, "How Dairy Farmers Manage the Interactions between Organic Fertilizers and Earthworm Ecotypes and Their Predators."

56. Ribot and Peluso, "A Theory of Access," 157.

57. Campling and Havice, "The Problem of Property in Industrial Fisheries."

58. Proby, "Pickers Stealing His Worms, Farmer Says."

59. Globe and Mail, "Beatings and Shotgun Fire Heating up Worm Wars."

60. Government of Canada, "Tillage and Seeding Practices, Census of Agriculture, 2021."

61. Briones and Schmidt, "Conventional Tillage Decreases the Abundance and Biomass of Earthworms and Alters Their Community Structure in a Global Meta-Analysis."

62. Fox et al., "Earthworm Population Dynamics as a Consequence of Long-Term and Recently Imposed Tillage in a Clay Loam Soil."

63. Miller et al., "Short-Term Legacy Effects of Feedlot Manure Amendments on Earthworm Abundance in a Clay Loam Soil"; Whalen, Parmelee, and Edwards, "Population Dynamics of Earthworm Communities in Corn Agroecosystems Receiving Organic or Inorganic Fertilizer Amendments"; Sharpley et al., "Land Application of Manure Can Influence Earthworm Activity and Soil Phosphorus Distribution."

CHAPTER 3

1. Marx, *Capital: Volume III*; Ball, "Differential Rent and the Role of Landed Property."

2. Mazzucato, *The Value of Everything*.

3. Baglioni, Campling, and Hanlon, "Beyond Rentiership," 1530.

4. Hudson, "Cultural Political Economy Meets Global Production Networks: A Productive Meeting?"

5. Harvey, *The Limits to Capital*, 360.

6. Vercellone, "From Formal Subsumption to General Intellect: Elements for a Marxist Reading of the Thesis of Cognitive Capitalism"

7. Goodman and Redclift, "Capitalism, Petty Commodity Production and the Farm Enterprise," 241.

8. R. Murray, "Value and Theory of Rent: Part One," 112.

9. R. Murray, "Value and Theory of Rent: Part One," 120–21.

10. Boyd and Prudham, "On the Themed Collection: The Formal and Real Subsumption of Nature."

11. Fine, "On Marx's Theory of Agricultural Rent."

12. Capps, "Tribal-Landed Property: The Value of the Chieftaincy in Contemporary Africa."

13. Keith et al., "Consequences of Anecic Earthworm Removal over 18 Months for Earthworm Assemblages and Nutrient Cycling in a Grassland."

14. Liu and Zou, "Exotic Earthworms Accelerate Plant Litter Decomposition in a Puerto Rican Pasture and a Wet Forest"; Milcu et al., "Earthworms and Legumes Control Litter Decomposition in a Plant Diversity Gradient."

15. Fischer et al., "How Do Earthworms, Soil Texture and Plant Composition Affect Infiltration along an Experimental Plant Diversity Gradient in Grassland?"

16. The conservative Mennonite faith acts as an impediment to land in another way. Worm-picking regions overlap with a disproportionately large number of conservative Mennonite farmers, who take the Sabbath day of rest quite seriously. While these farmers will rent their land to pickers, they carefully stipulate that no one can work on their land between 12:00 a.m. Saturday night and 12:00 a.m. Sunday night, effectively eliminating two picking days a week.

17. Most farmers submit Nutrient Management Plans to OMAFRA to comply with provincial regulations to limit overspreading, which can cause significant environmental and human health problems.

18. Information from OMAFRA ("Number of Milk Producers by County"), estimated value and rental rate of farmland by county and township.

19. Dhiman and Satter, "Yield Response of Dairy Cows Fed Different Proportions of Alfalfa Silage and Corn Silage"; Brito et al., "Alfalfa Cut at Sundown and Harvested as Baleage Improves Milk Yield of Late-Lactation Dairy Cows"; Ferraretto and Shaver, "Meta-Analysis: Effect of Corn Silage Harvest Practices on Intake, Digestion, and Milk Production by Dairy Cows."

20. Jennings, *Alfalfa for Dairy Cattle*, 1.

21. Hoshmand, *Design of Experiments for Agriculture and the Natural Sciences*; Bockheim and Gennadiyev, "The Value of Controlled Experiments in Studying Soil-Forming Processes: A Review."

22. Engel-Di Mauro, *Ecology, Soils, and the Left: An Ecosocial Approach.*

CHAPTER 4

1. Portes and Sassen-Koob, "Making It Underground: Comparative Material on the Informal Sector in Western Market Economies," 31.

2. Hart, "Informal Income Opportunities and Urban Employment in Ghana."

3. Portes and Sassen-Koob, "Making It Underground: Comparative Material on the Informal Sector in Western Market Economies"; Basole and Basu, "Relations of Production and Modes of Surplus Extraction in India."

4. Portes and Sassen-Koob, "Making It Underground: Comparative Material on the Informal Sector in Western Market Economies," 31.

5. Portes and Haller, "The Informal Economy."

6. Hardt, "The Post-Operaist Approach to the Formal and Real Subsumption of Labor Under Capital."

7. Wilson, "Precarization, Informalization, and Marx"; Federici, "Notes on Gender in Marx's *Capital*"; Mezzadri, "A Value Theory of Inclusion."

8. Marx, *Capital: Volume I*, 412n54.

9. Committee Documents: Standing Committee on General Government—2006-Oct-25—Bill 148, Highway Traffic Amendment Act, Legislative Assembly of Ontario.

10. Parker, "Southwestern Ontario Land Values: 2020 Edition."

11. Giannaris, "Method and Medium for Coloring Live Bait Worms."

12. Davis, "The Worm Has Turned—Green."

13. Lavigne, "'Black Market' Unearthed Theft Part of War for Worms."

14. Volgenau, "Worm Wars."

15. Appleby, "Second Spill Reveals Rough Trade: Business in Worm Capital of the World Sparks Turf Wars."

16. R V Tran, No. 5747 (Ontario Court of Justice).

17. Canada Revenue Agency, "Toronto Resident Sentenced to Five Years in Prison for GST/HST Fraud."

18. None of this is new to the CRA, which continues to monitor the industry largely through the coolers, whose fixed capital and permanency provide an entry point for investigations. Apollo Baits receives letters from the CRA "telling me I have to garnish paid amounts" to some of their crew chiefs. Another cooler mentioned to me: "They watch the business. They watch everyone. They know."

CHAPTER 5

Portions of this chapter were previously published in Steckley, "'Completely Free': How a Subsumption of Labour and Nature Framework Explains the Surprising Expressions of 'Freedom' by Immigrant Worm Pickers in Ontario."

1. Marx, *Capital: Volume I*, 1021.
2. Marx, *Capital: Volume I*, 1021.
3. *Montreal Star*, "Worm Picking Big Business Now."
4. Tomlin, "The Earthworm Bait Market in North America."
5. Decision No. 5881/89.
6. Marx, *Capital: Volume I*, 694.
7. Freeman and Kleiner, "The Last American Shoe Manufacturers: Decreasing Productivity and Increasing Profits in the Shift from Piece Rates to Continuous Flow Production," 308.
8. Bender, Green, and Heywood, "Piece Rates and Workplace Injury: Does Survey Evidence Support Adam Smith?"; C. Smith, "The Double Indeterminacy of Labour Power: Labour Effort and Labour Mobility."
9. Baglioni, "Labour Control and the Labour Question in Global Production Networks"; Gidwani, "The Cultural Logic of Work: Explaining Labour Deployment and Piece-Rate Contracts in Matar Taluka, Gujarat"; Ekers and Farnan, "Planting the Nation: Tree Planting Art and the Endurance of Canadian Nationalism"; Smith, "The Double Indeterminacy of Labour Power."
10. Breman, *Footloose Labour: Working in India's Informal Economy*, 2:239.
11. Gidwani, "The Cultural Logic of Work: Explaining Labour Deployment and Piece-Rate Contracts in Matar Taluka, Gujarat," 94.
12. Brown, "Firms' Choice of Method of Pay."
13. Freeman and Kleiner, "The Last American Shoe Manufacturers: Decreasing Productivity and Increasing Profits in the Shift from Piece Rates to Continuous Flow Production."
14. Piece rates often have the added benefit of undercutting worker solidarity, a feature across industries. As Ekers and Farnan note for tree planting, many middle-class college kids are skeptical of what benefits unions would bring them considering their background (likely governmental and parental benefits that would otherwise be provided by a union); see "Planting the Nation: Tree Planting Art and the Endurance of Canadian Nationalism." But more vulnerable workers may also prefer the individualized pay structure. Non-status or migrant workers may prefer piece-rate wages as they are already excluded from claiming state or union benefits, or may

simply want to make as much money in a short as possible to time to avoid a government crackdown on illegal immigration; see Guthman, *Wilted: Pathogens, Chemicals, and the Fragile Future of the Strawberry Industry*.

15. Katz and Monk, "When in the World Are Women?"; Orzeck, "What Does Not Kill You: Historical Materialism and the Body."

16. Taylor and Rioux, *Global Labour Studies*.

17. Melamed, "Racial Capitalism."

18. De Genova, "A Racial Theory of Labour."

19. Federici, "Notes on Gender in Marx's *Capital*"; Harvey, "The Body as an Accumulation Strategy"; Wright, *Disposable Women and Other Myths of Global Capitalism*.

20. Mills, "From Nimble Fingers to Raised Fists: Women and Labor Activism in Globalizing Thailand."

21. Elson and Pearson, "'Nimble Fingers Make Cheap Workers': An Analysis of Women's Employment in Third World Export Manufacturing."

22. Glover and Guerrier, "Women in Professional IT Jobs in the UK: Old Wine in New Bottles?"

23. Mills, "From Nimble Fingers to Raised Fists: Women and Labor Activism in Globalizing Thailand"; Vosko and Casey, "Enforcing Employment Standards for Temporary Migrant Agricultural Workers in Ontario, Canada: Exposing Underexplored Layers of Vulnerability." Nor are female worm pickers more "obedient," "passive," or easier to discipline. As I will demonstrate below, the demand for labor and conditions of work allows considerable worker control over the labor process. No interviewee, male or female, described physical abuse from their employers (who were rarely present on the field anyway). This does not mean abuse did not, or does not occur, only that interviewees told me that it neither has happened to them nor have they witnessed it. I must note, however, that the limited sample size of worm pickers combined with my positionality as a white, male Canadian researcher conducting interviews through an interpreter significantly impedes the sharing of such individualized and intimate experience.

24. In relation to nursing, see Musshauser et al., "The Impact of Sociodemographic Factors vs. Gender Roles on Female Hospital Workers' Health: Do We Need to Shift Emphasis?"; Silva-Costa et al., "Need for Recovery from Work and Sleep-Related Complaints among Nursing Professionals."

25. Bihan and Martin, "Atypical Working Hours: Consequences for Childcare Arrangements."

26. Lanphier, "Canada's Response to Refugees."

27. Boyd, Perron, and Cowan, "The Contemporary Labor Market Integration of Vietnamese Refugees in Canada," 5.

28. M. Boyd, "Visible Minority and Immigrant Earnings Inequality: Reassessing an Employment Equity Premise"; Pfeifer, "Community, Adaptation, and the Vietnamese in Toronto"; Hagey et al., "Immigrant Nurses' Experience of Racism."

29. Statistics Canada, "Minimum Wage in Canada since 1975."

30. Pfeifer, "Community, Adaptation, and the Vietnamese in Toronto."

31. No picker I spoke with felt any social stigma attached to worm picking. This contrasts with reports of earlier Portuguese immigrants. The teenage kids of some worm pickers would plead with their parents "to be dropped off a block away, so the other school kids would not tease them about being worm pickers"; see Blank, "A Story of Portuguese Immigrants in Toronto and a Yellow School Bus." 11.

32. UNHCR, "Refugees in Canada."

33. Houle, "Results from the 2016 Census: Syrian Refugees Who Resettled in Canada in 2015 and 2016."

34. UNHCR, "Seven Decades of Refugee Protection in Canada: 1950–2020."

35. Puzic, "Fact Check: Do Refugees Get More Financial Help Than Canadian Pensioners?"; Mas, "Do Government-Assisted Refugees Receive More Money for Food than Canadians on Welfare?"

36. Government of Canada, "Guide 2201—Community Sponsors to Privately Sponsor Refugees."

37. Statistics Canada, "Minimum Wage in Canada since 1975."

CHAPTER 6

Second epigraph: Michael Lunn, as told to the *Star Tribune*. Used with permission. Original caption for figure 13: Worm Picking Big Business Now. A professional worm picker in Toronto can pick close to 2,500 worms an hour. Working at night on golf courses, the pickers use head-lamps and pails strapped to their legs. Close to 12,000 worms a night are taken from Toronto's metropolitan courses. Date published: Tues., June 16, 1959, page 55. Credit: *Montreal Star*. Portions of this chapter were previously published in Steckley, "'Completely Free': How a Subsumption of Labour and Nature Framework Explains the Surprising Expressions of 'Freedom' by Immigrant Worm Pickers in Ontario."

1. Tsing, *The Mushroom at the End of the World: On the Possibility of Life in Capitalist Ruins*, 77.

2. Banaji, "The Fictions of Free Labour: Contract, Coercion, and So-Called Unfree Labour."

3. Lerche, "The Unfree Labour Category and Unfree Labour Estimates: A Continuum within Low-End Labour Relations"; Rogaly, "Migrant Workers in the ILO's Global Alliance Against Forced Labour Report: A Critical Appraisal."

4. McGrath, "Unfree Labor."

5. LeBaron, "Unfree Labour beyond Binaries: Insecurity, Social Hierarchy and Labour Market Restructuring."

6. Barrientos, Kothari, and Phillips, "Dynamics of Unfree Labour in the Contemporary Global Economy."

7. D. Mitchell, "Labor's Geography: Capital, Violence, Guest Workers and the Post-World War II Landscape."

8. Binford, "Assessing Temporary Foreign Worker Programs through the Prism of Canada's Seasonal Agricultural Worker Program: Can They Be Reformed or Should They Be Eliminated?"

9. McGrath, "Unfree Labor," 1016.

10. Calvao, "Unfree Labor," 458.

11. Castree, "Labour Geography: A Work in Progress"; Coe and Jordhus-Lier, "Constrained Agency? Re-Evaluating the Geographies of Labour"; Coe and Jordhus-Lier, "The Multiple Geographies of Constrained Labour Agency"; Herod, "From a Geography of Labor to a Labor Geography: Labor's Spatial Fix and the Geography of Capitalism"; Herod, "Labour Geography: Where Have We Been? Where Should We Go?"

12. Coe and Jordhus-Lier, "The Multiple Geographies of Constrained Labour Agency."

13. Strauss, "Labour Geography III: Precarity, Racial Capitalisms and Infrastructure"; McDowell, Batnitzky, and Dyer, "Division, Segmentation, and Interpellation: The Embodied Labors of Migrant Workers in a Greater London Hotel"; Maldonado, "'It Is Their Nature to Do Menial Labour': The Racialization of 'Latino/a Workers' by Agricultural Employers"; Reid-Musson, "Grown Close to Home": Migrant Farmworker (Im) Mobilities and Unfreedom on Canadian Family Farms"; Preibisch, "Pick-Your-Own Labor: Migrant Workers and Flexibility in Canadian Agriculture 1"; Baglioni, "Labour Control and the Labour Question in Global Production Networks"; Fudge, "(Re) Conceptualising Unfree Labour: Local Labour Control Regimes and Constraints on Workers' Freedoms."

14. D. Mitchell, "Dead Labor and the Political Economy of Landscape—California Living, California Dying."

15. Bakker, "Neoliberalizing Nature? Market Environmentalism in Water Supply in England and Wales"; Bakker and Bridge, "Material Worlds? Resource Geographies and the Matter of Nature'"; Collard and Dempsey, "Life for Sale? The Politics of Lively Commodities."

16. Baglioni and Campling, "Natural Resource Industries as Global Value Chains: Frontiers, Fetishism, Labour and the State."

17. Mishra, "Urbanisation Through Brick Kilns: The Interrelationship Between Appropriation of Nature and Labour Regimes."

18. Velásquez and Ayala, "Production of Nature and Labour Agency: How the Subsumption of Nature Affects Trade Union Action in the Fishery And Aquaculture Sectors in Aysén, Chile."

19. Marx, *Capital: Volume I*, 506.

20. Tomba, "On the Capitalist and Emancipatory Use of Asynchronies in Formal Subsumption," 293, 287.

21. Vercellone, "From Formal Subsumption to General Intellect: Elements for a Marxist Reading of the Thesis of Cognitive Capitalism," 20.

22. Joyce, "Rediscovering the Cash Nexus, Again: Subsumption and the Labour–Capital Relation in Platform Work."

23. Guthman, *Wilted: Pathogens, Chemicals, and the Fragile Future of the Strawberry Industry*.

24. C. Smith, "The Double Indeterminacy of Labour Power," 393.

25. Rogaly, "Migrant Workers in the ILO's Global Alliance Against Forced Labour Report: A Critical Appraisal."

26. Tsing, "Free in the Forest: Popular Neoliberalism and the Aftermath of War in the US Pacific Northwest."

27. Le v. Canada, No. 1263 (Minister of Citizenship and Immigration).

28. Campbell et al., "A Comparison of Health Access between Permanent Residents, Undocumented Immigrants and Refugee Claimants in Toronto, Canada," 165.

29. Magalhaes, Carrasco, and Gastaldo, "Undocumented Migrants in Canada: A Scope Literature Review on Health, Access to Services, and Working Conditions."

30. D. Mitchell, *They Saved the Crops: Labor, Landscape, and the Struggle over Industrial Farming in Bracero-Era California*.

31. Shaer, "Migrant Worker Programs a Win-Win for Workers and Farmers."

32. Vosko and Casey, "Enforcing Employment Standards for Temporary Migrant Agricultural Workers in Ontario, Canada: Exposing Underexplored Layers of Vulnerability."

33. Gabriel and Macdonald, "After the International Organization for Migration: Recruitment of Guatemalan Temporary Agricultural Workers to Canada"; Foster and Taylor, "In the Shadows: Exploring the Notion of 'Community' for Temporary Foreign Workers in a Boomtown."

34. Reid-Musson, "Grown Close to Home™: Migrant Farmworker (Im) Mobilities and Unfreedom on Canadian Family Farms."

35. Strauss and McGrath, "Temporary Migration, Precarious Employment and Unfree Labour Relations: Exploring the 'Continuum of Exploitation' in Canada's Temporary Foreign Worker Program."

36. Choudry and Smith, *Unfree Labour?: Struggles of Migrant and Immigrant Workers in Canada*.

37. Fudge, "(Re) Conceptualising Unfree Labour: Local Labour Control Regimes and Constraints on Workers' Freedoms."

38. Martin, "Migrant Workers in Commercial Agriculture."

39. Ferguson, "Conditions Tough for Canada's Migrant Workers."

40. V. Kirsch, "Migrant Workers Decry 'Slavery.'"

41. Monteiro, "Mexican Couple Fear Death If They Go Home Again; Pair Who Complained about Working Conditions at Worm-Picking Operation in Guelph Seeking Refugee Status."

42. In agriculture, employers often provide housing (or bunkhouses) on-site, using portables, trailers, or dwellings constructed purposely for TFWs. When providing room and board directly, employers have a right to charge employees up to $85/week. Obviously, this can lead to situations where employers can profit from crowded accommodations, shoddy construction, and a general lack of amenities. For

more on such conditions as they relate to the TFWP in Ontario, see Fudge and McPhail, "The Temporary Foreign Worker Program in Canada: Low-Skilled Workers as an Extreme Form of Flexible Labour."

CHAPTER 7

1. OMAFRA, "Number of Milk Producers by County."

2. For a more detailed story of this account and how it relates to Canadian nightcrawlers, see Amy Stewart's retelling in *The Earth Moved*, 99–109.

3. Keller et al., "From Bait Shops to the Forest Floor: Earthworm Use and Disposal by Anglers."

4. Department of Natural Resources, "Minnesota Fishing Regulations."

5. Friends of Algonquin Park, "Special Fishing Regulations in Algonquin Park—Summary."

6. Animal and Plant Health Inspection Service, "Earthworms."

7. USA Trade Online.

8. Goodman, Sorj, and Wilkinson, *From Farming to Biotechnology: A Theory of Agro-Industrial Development*.

9. Berkley, "Trust Facts, Not Fads. Trust Powerbait."

10. For more on the differences between live and artificial bait in practical settings, see Clapp and Clark, "Hooking Mortality of Smallmouth Bass Caught on Live Minnows and Artificial Spinners"; Siewert and Cave, "Survival of Released Bluegill, Lepomis Macrochirus, Caught on Artificial Flies, Worms, and Spinner Lures"; Arlinghaus et al., "Size Selectivity, Injury, Handling Time, and Determinants of Initial Hooking Mortality in Recreational Angling for Northern Pike: The Influence of Type and Size of Bait"; Bailey et al., "Live versus Artificial Bait Influences on Walleye (Sander Vitreus) Angler Effort and Catch Rates on Escanaba Lake, Wisconsin, 1993–2015."

11. Bailey et al., "Live versus Artificial Bait Influences on Walleye (Sander Vitreus) Angler Effort and Catch Rates on Escanaba Lake, Wisconsin, 1993–2015."

12. Vincent et al., "Changes in Canada's Climate: Trends in Indices Based on Daily Temperature and Precipitation Data," 336.

13. Dundas and Haefen, "The Effects of Weather on Recreational Fishing Demand and Adaptation: Implications for a Changing Climate."

14. Hunt and Moore, "The Potential Impacts of Climate Change on Recreational Fishing in Northern Ontario."

15. Ficke, Myrick, and Hansen, "Potential Impacts of Global Climate Change on Freshwater Fisheries"; Creighton et al., "Climate Change and Recreational Fishing: Implications of Climate Change for Recreational Fishers and the Recreational Fishing Industry."

16. Jones et al., "Climate Change Impacts on Freshwater Recreational Fishing in the United States."

17. Jones et al., "Climate Change Impacts on Freshwater Recreational Fishing in the United States." See also Hunt et al., "Identifying Alternate Pathways for Climate Change to Impact Inland Recreational Fishers."

18. Anderson et al., "On Assemblages and Geography."

CONCLUSION

1. Thompson, *Whigs and Hunters: The Origin of the Black Acts*.

2. Barua, "Animating Capital: Work, Commodities, Circulation," 665, 655.

3. C. Smith, "The Double Indeterminacy of Labour Power."

4. Baglioni and Campling, "Natural Resource Industries as Global Value Chains: Frontiers, Fetishism, Labour and the State."

5. Bakker, "Neoliberalizing Nature? Market Environmentalism in Water Supply in England and Wales"; Barua, "Animating Capital: Work, Commodities, Circulation"; Collard and Dempsey, "Life for Sale? The Politics of Lively Commodities."

6. Kautsky, *The Agrarian Question: In Two Volumes*.

7. Lowe and Butt, "Culture Techniques for Soil Dwelling Earthworms: A Review."

8. Lowe, Butt, and Sherman, "Current and Potential Benefits of Mass Earthworm Culture."

9. Logt et al., "The Anecic Earthworm Lumbricus Terrestris Can Persist after Introduction into Permanent Grassland on Sandy Soil," 6.

10. For a lovely exception to this see Thammavongsa, *How to Pronounce Knife: Stories*. This collection includes a fictionalized story centered around worm picking.

11. Kennedy, "Canada's Migrant Farm Worker Program Was Founded on 'Racist' Policies, New Lawsuit Alleges. And Today's Workers Are Still Paying the Price."

12. Miller, "The Golf Course Worm Is an Educated Worm."

BIBLIOGRAPHY

Abel, A. "Subterranean Harvest." *Canadian Geographic*, September–October 1998, 28.

Aberdeen Daily News. "Goin' a-Fishin'? How to Get and Keep the Bait that Gets the Fish." July 22, 1905, 2.

Akram-Lodhi, A. H., and C. Kay. "Surveying the Agrarian Question (Part 2): Current Debates and Beyond." *The Journal of Peasant Studies* 37, no. 2 (2010): 255–84.

Anderson, B., M. Kearnes, C. McFarlane, and D. Swanton. "On Assemblages and Geography." *Dialogues in Human Geography* 2, no. 2 (2012): 171–89.

Anderson, B., and C. McFarlane. "Assemblage and Geography." *Area* 43, no. 2 (2011): 124–27.

Andriuzzi, W. S., M. M. Pulleman, O. Schmidt, J. H. Faber, and L. Brussaard. "Anecic Earthworms (Lumbricus terrestris) Alleviate Negative Effects of Extreme Rainfall Events on Soil and Plants in Field Mesocosms." *Plant and Soil* 397, no. 1 (2015): 103–13.

Animal and Plant Health Inspection Service. "Earthworms." United States Department of Agriculture, June 17, 2022.

Appleby, T. "Second Spill Reveals Rough Trade: Business in Worm Capital of the World Sparks Turf Wars." *Globe and Mail*, May 25, 1993, 16.

Arlinghaus, R., T. Klefoth, A. Kobler, and S. J. Cooke. "Size Selectivity, Injury, Handling Time, and Determinants of Initial Hooking Mortality in Recreational Angling for Northern Pike: The Influence of Type and Size of Bait." *North American Journal of Fisheries Management* 28, no. 1 (2008): 123–34.

Baden, W. W., and C. S. Beekman. "Culture and Agriculture: A Comment on Sissel Schroeder, *Maize Productivity in the Eastern Woodlands and Great Plains of North America*." *American Antiquity* 66, no. 3 (2001): 505–15.

Baglioni, E. "Labour Control and the Labour Question in Global Production Networks: Exploitation and Disciplining In Senegalese Export Horticulture." *Journal of Economic Geography* 18, no. 1 (2018): 111–37.

Baglioni, E., and L. Campling. "Natural Resource Industries as Global Value Chains: Frontiers, Fetishism, Labour and the State." *Environment and Planning A: Economy and Space* 49, no. 11 (2017): 2437–56.

Baglioni, E., L. Campling, and G. Hanlon. "Beyond Rentiership: Standardisation, Intangibles and Value Capture in Global Production." *Environment and Planning A: Economy and Space* 55, no. 6 (2023): 1528–47.

Bailey, C. T., A. M. Noring, S. L. Shaw, and G. G. Sass. "Live Versus Artificial Bait Influences on Walleye (Sander Vitreus) Angler Effort and Catch Rates on Escanaba Lake, Wisconsin, 1993–2015." *Fisheries Research* 219 (2019): 105330.

Bain, A., and G. King. "Asylum for Wayward Immigrants: Historic Ports and Colonial Settlements In Northeast North America." *Journal of the North Atlantic* 2, sp. 1 (2011): 109–24.

Bakker, K. "Neoliberalizing Nature? Market Environmentalism in Water Supply in England and Wales." *Annals of the Association of American Geographers* 95, no. 3 (2005): 542–65.

Bakker, K., and G. Bridge. "Material Worlds? Resource Geographies and the Matter of Nature." *Progress in Human Geography* 30, no. 1 (2006): 5–27.

Ball, M. "Differential Rent and the Role of Landed Property." *International Journal of Urban and Regional Research* 1, no. 1–3 (1977): 380–403.

Banaji, J. "The Fictions of Free Labour: Contract, Coercion, and So-Called Unfree Labour." *Historical Materialism* 11, no. 3 (2003): 69–95.

Barrientos, S., U. Kothari, and N. Phillips. "Dynamics of Unfree Labour in the Contemporary Global Economy." *The Journal of Development Studies* 49, no. 8 (2013): 1037–41.

Barua, M. "Animating Capital: Work, Commodities, Circulation." *Progress in Human Geography* 43, no. 4 (2019): 650–69.

Basole, A., and D. Basu. "Relations of Production and Modes of Surplus Extraction in India: Part I: Agriculture." *Economic and Political Weekly* (2011): 41–58.

Bellon, T. "In Roundup Case, U.S. Judge Cuts $2 Billion Verdict Against Bayer to $86 Million." Reuters, July 25, 2019. https://www.reuters.com/article/us-bayer -glyphosate-lawsuit-idUSKCN1UL03G.

Beltaire, M. "Running a Golf Course Is No Picnic." *Detroit Free Press*, September 15, 1953, 46.

Bender, K. A., C. P. Green, and J. S. Heywood. "Piece Rates and Workplace Injury: Does Survey Evidence Support Adam Smith?" *Journal of Population Economics* 25, no. 2 (2012): 569–90.

Bennett, J. *Vibrant Matter: A Political Ecology of Things.* Duke University Press, 2010.

Berkley. "Trust Facts, Not Fads. Trust Powerbait." 2022. https://www.berkley -fishing.com/pages/berkley-powerbait.

Berners, J. *An Older Form of The Treatyse of Fysshynge Wyth an Angle* (Issue 9). W. Satchell, 1883.

Bernstein, H. "Is There an Agrarian Question in the 21st Century?" *Canadian Journal of Development Studies/Revue Canadienne d'études Du Développement* 27, no. 4 (2006): 449–60.

Bertrand, M., S. Barot, M. Blouin, J. Whalen, T. Oliveira, and J. Roger-Estrade. "Earthworm Services for Cropping Systems. A Review." *Agronomy for Sustainable Development* 35, no. 2 (2015): 553–67.

Bihan, B.L., and C. Martin. "Atypical Working Hours: Consequences for Childcare Arrangements." *Social Policy & Administration* 38, no. 6 (2004): 565–90.

Binet, F., and P. Trehen. "Experimental Microcosm Study of the Role of Lumbricus Terrestris (Oligochaeta: Lumbricidae) on Nitrogen Dynamics in Cultivated Soils." *Soil Biology and Biochemistry* 24, no. 12 (1992): 1501–06.

Binford, A.L. "Assessing Temporary Foreign Worker Programs through the Prism of Canada's Seasonal Agricultural Worker Program: Can They Be Reformed or Should They Be Eliminated?" *Dialectical Anthropology* 43, no. 4 (2019): 347–66.

Birkás, M., M. Jolánkai, C. Gyuricza, and A. Percze. "Tillage Effects on Compaction, Earthworms and Other Soil Quality Indicators in Hungary." *Soil and Tillage Research* 78, no. 2 (2004): 185–96.

Blank, R. "A Story of Portuguese Immigrants in Toronto and a Yellow School Bus." Canadian Centre for Azorean Research and Studies, 2015. http://ccars.apps01 .yorku.ca/wp-content/uploads/2017/09/ApanharMinhoca.pdf.

Bockheim, J.G., and A.N. Gennadiyev. "The Value of Controlled Experiments in Studying Soil-Forming Processes: A Review." *Geoderma* 152, no. 3-4 (2009): 208–17.

Borron, E.B. "Report on the Basin of Moose River and Adjacent Country belonging to the Province of Ontario" (No. 87; Sessional Papers). Ontario Legislative Assembly, 1890.

Boston Daily Globe. "Makes Money Selling Worms." October 3, 1926, B48.

Bouché, M.B. "Strategies lombriciennes." *Ecological Bulletin (Stockholm)* 24 (1977): 122–32.

Boyd, M. "Immigration Policies and Trends: A Comparison of Canada and the United States." *Demography* 13, no. 1 (1976): 83–104.

———. "Gender, Visible Minority, and Immigrant Earnings Inequality: Reassessing an Employment Equity Premise." In *Deconstructing a Nation: Immigration, Multiculturalism, and Racism in '90s Canada*, edited by V. Satzewich, 279–322. Fernwood Publishing, 1992.

Boyd, M., S. Perron, and D.G. Cowan. "The Contemporary Labor Market Integration of Vietnamese Refugees in Canada." *Journal of Immigrant & Refugee Studies* (2021): 1–18.

Boyd, W. "Making Meat: Science, Technology, and American Poultry Production." *Technology and Culture* 42, no. 4 (2001): 631–64.

Boyd, W., and S. Prudham. "On the Themed Collection: The Formal and Real Subsumption of Nature." *Society & Natural Resources* 30, no. 7 (2017): 877–84.

Boyd, W., W.S. Prudham, and R.A. Schurman. "Industrial Dynamics and the Problem of Nature." *Society & Natural Resources* 14, no. 7 (2001): 555–70.

Braun, B. "The 2013 Antipode RGS-IBG Lecture: New Materialisms and Neoliberal Natures." *Antipode* 47, no. 1 (2015): 1–14.

Breman, J. *Footloose Labour: Working in India's Informal Economy.* Cambridge University Press, 1996.

Briones, M. J. I., and O. Schmidt. "Conventional Tillage Decreases the Abundance and Biomass of Earthworms and Alters Their Community Structure in a Global Meta-Analysis." *Global Change Biology* 23, no. 10 (2017): 4396–4419.

Brito, A. F., G. F. Tremblay, A. Bertrand, Y. Castonguay, G. Bélanger, R. Michaud, and R. Berthiaume. "Alfalfa Cut at Sundown and Harvested as Baleage Improves Milk Yield of Late-Lactation Dairy Cows." *Journal of Dairy Science* 91, no. 10 (2008): 3968–82.

Broomhall, P. "Can't Beat Those Worms." *Vancouver Sun*, March 29, 1957, 26.

Brown, C. "Firms' Choice of Method of Pay." *ILR Review* 43, no. 3 (1990): 165–82.

Bryant, N. "Anglers Still Favor Earthworms." *New York Times*, 1978, 323.

———. "A Case for Fishing with Worms." *New York Times*, July 9, 1984, 11.

Burkett, P. "Value, Capital and Nature: Some Ecological Implications of Marx's Critique of Political Economy." *Science & Society* (1996): 332–59.

Butt, K. R. "Food Quality Affects Production of Lumbricus Terrestris (L.) under Controlled Environmental Conditions." *Soil Biology and Biochemistry* 43, no. 10 (2011): 2169–75.

Butt, K. R., and C. N. Lowe. "Controlled Cultivation of Endogeic and Anecic Earthworms." In *Biology of Earthworms*, edited by A. Karaca, 107–110. Springer, 2011.

Buxton, R. "Worm Fishing for Trout." In *Turf, Field, and Farm*. New York, 1867.

Callaham, M., and P. Hendrix. "Relative Abundance and Seasonal Activity of Earthworms (Lumbricidae and Megascolecidae) as Determined by Hand-Sorting and Formalin Extraction in Forest Soils on the Southern Appalachian Piedmont." *Soil Biology and Biochemistry* 29, no. 3–4 (1997): 317–21.

Callon, M. "Economic Markets and the Rise of Interactive Agencements: From Prosthetic Agencies to Habilitated Agencies." In *Living in a Material World: Economic Sociology Meets Science and Technology Studies*, edited by T. Pinch and R. Swedberg, 29–56. MIT Press, 2008.

Calvao, F. "Unfree Labor." *Annual Review of Anthropology* 45, no. 1 (2016): 451–67.

Camp, R. "Albert, Fly Purist Take 4 Brown Trout, But His Heart Is Heavy, He Used Worms." *The New York Times*, April 15, 1951, 147.

Campbell, R. M., A. G. Klei, B. D. Hodges, D. Fisman, and S. Kitto. "A Comparison of Health Access between Permanent Residents, Undocumented Immigrants and Refugee Claimants in Toronto, Canada." *Journal of Immigrant and Minority Health* 16, no. 1 (2014): 165–76.

Campling, L., and E. Havice. "The Problem of Property in Industrial Fisheries." *The Journal of Peasant Studies* 41, no. 5 (2014): 707–27.

Canada Revenue Agency. "Toronto Resident Sentenced to Five Years in Prison for GST/HST Fraud." 2017. https://www.canada.ca/en/revenue-agency/news/newsroom/criminal-investigations-actions-charges-convictions/toronto-resident-sentenced-gsthst-fraud-20171117.html.

Canadian Soil Information Service. *Soil Landscapes of Canada.* 2019. http://sis.agr.gc.ca.

Capowiez, Y., S. Sammartino, and E. Michel. "Burrow Systems of Endogeic Earthworms: Effects of Earthworm Abundance and Consequences for Soil Water Infiltration." *Pedobiologia* 57, no. 4–6 (2014): 303–9.

Capps, G. "Tribal-Landed Property: The Value of the Chieftaincy in Contemporary Africa." *Journal of Agrarian Change* 16, no. 3 (2016): 452–77.

Carson, R. *Silent Spring*. Houghton Mifflin Harcourt, 2002.

Carton, W., and E. Andersson. "Where Forest Carbon Meets Its Maker: Forestry-Based Offsetting as the Subsumption of Nature." *Society & Natural Resources* 30, no. 7 (2017): 829–43.

Castree, N. "False Antitheses? Marxism, Nature and Actor-Networks." *Antipode* 34, no. 1 (2002): 111–46.

———. "Labour Geography: A Work in Progress." *International Journal of Urban and Regional Research* 31, no. 4 (2007): 853–62.

Champ, D. "Nightcrawlers Make Good Bait in Warm Water." *Sioux City Journal* (Iowa), 2002. http://siouxcityjournal.com/sports/recreation/outdoors/nightcrawlers-make-good-bait-in-warm-water/article_c62d4562-80b4-5bc1-9928-8816e8f320cd.html

Cherry, Z. "After a Fashion: Night Creature Not Men from Mars—Just Worm Pickers." *Globe and Mail*, May 20, 1968, 16.

Choudry, A., and A. A. Smith. *Unfree Labour?: Struggles of Migrant and Immigrant Workers in Canada*. PM Press, 2016.

Clapp, D. F., and R. D. Clark, Jr. "Hooking Mortality of Smallmouth Bass Caught on Live Minnows and Artificial Spinners." *North American Journal of Fisheries Management* 9, no. 1 (1989): 81–85.

Coe, N. M., and D. C. Jordhus-Lier. "Constrained Agency? Re-Evaluating the Geographies of Labour." *Progress in Human Geography* 35, no. 2 (2011): 211–33.

———. "The Multiple Geographies of Constrained Labour Agency." *Progress in Human Geography* 47, no. 4 (2023): 533–54.

Collard, R.-C., and J. Dempsey. "Life for Sale? The Politics of Lively Commodities." *Environment and Planning A* 45, no. 11 (2013): 2682–99.

Committee Documents: Standing Committee on General Government—2006-Oct-25—Bill 148, Highway Traffic Amendment Act, No. 148, Legislative Assembly of Ontario (2006). http://www.ontla.on.ca/web/committee-proceedings/committee_transcripts_details.do?locale=en&BillID=473&ParlCommID=7422&Business=148&Date=2006-10-25&DocumentID=20816.

Cook, I., D. Crouch, S. Naylor, and J. Ryan. *Cultural Turns/Geographical Turns: Perspectives on Cultural Geography*. Routledge, 2014.

Cooper, C. "There's a Gold Mine in Worms." *Austin American-Statesman*, August 29, 1979, 12.

Cooper, M. H. "Open up and Say 'Baa': Examining the Stomachs of Ruminant Livestock and the Real Subsumption of Nature." *Society & Natural Resources* 30, no. 7 (2017): 812–28.

Creighton, C., B. Sawynok, S. Sutton, D. D'Silva, I. Stagles, C. Pam, and D. Spooner. "Climate Change and Recreational Fishing: Implications of Climate Change for

Recreational Fishers and the Recreational Fishing Industry." Fisheries Research and Development Corporation, 2013.

Crosson, A. "If You like Fishing, Prepare to Pay a Little More for Your Worms This Year." Public Radio International, 2014. https://www.pri.org/stories/2014-07-16/if-you-fishing-prepare-pay-little-more-your-worms-year.

"Current Land Claims." (n.d.). Retrieved July 8, 2024, from http://www.ontario.ca/page/current-land-claims.

Curry, J. P. "Factors Affecting the Abundance of Earthworms in Soils." *Earthworm Ecology* 9 (2004): 113.

Daily Spectrum (St. George, Utah). "Worm-Pickers Find a Buried Treasure." August 21, 1983, 21.

Darwin, C. *The Formation of Vegetable Mould, Through the Action of Worms: With Observations on Their Habits*. London, 1882.

Davis, G. "The Worm Has Turned—Green." *LA Times*, July 26, 2000. https://www.latimes.com/archives/la-xpm-2000-jul-26-me-59477-story.html.

De Genova, N. "A Racial Theory of Labour: Racial Capitalism from Colonial Slavery to Postcolonial Migration." *Historical Materialism* 1 (2023): 1–33.

De Grazia, S. *Of Time, Work, and Leisure*. New York: Twentieth Century Fund, 1962.

Decision No. 5881/89. Ontario Workplace Safety and Insurance Appeals Tribunal, 1991.

Department of Natural Resources. "Minnesota Fishing Regulations." 2021. https://files.dnr.state.mn.us/rlp/regulations/fishing/fishing_regs.pdf.

Dhiman, T. R., and L. D. Satter. "Yield Response of Dairy Cows Fed Different Proportions of Alfalfa Silage and Corn Silage." *Journal of Dairy Science* 80, no. 9 (1997): 2069–82.

Dittbrenner, N., H. Schmitt, and Y. Capowiez. "Sensitivity of Eisenia Fetida in Comparison to Aporrectodea Caliginosa and Lumbricus Terrestris after Imidacloprid Exposure." *Body Mass Change and Histopathology. J Soils Sediments* 11 (2011): 1000.

Dominguez, J. "State-of-the-Art and New Perspectives on Vermicomposting Research." In *Earthworm Ecology*, edited by C. A. Edwards, 401–24. CRC Press, 2004.

Dominguez, J., and C. A. Edwards. "Biology and Ecology of Earthworm Species Used for Vermicomposting." In *Vermiculture Technology: Earthworms, Organic Waste and Environmental Management*, edited by C. A. Edwards, N. Q. Arancon, and R. Sherman, 27–40. CRC Press, 2011.

Doolittle, W. E. "Permanent vs. Shifting Cultivation in the Eastern Woodlands of North America Prior to European Contact." *Agriculture and Human Values* 21, no. 2 (2004): 181–89.

Duluth New Tribune. "The Humble Angleworm." May 23, 1896, 6.

Dundas, S. J., and R. H. Haefen. "The Effects of Weather on Recreational Fishing Demand and Adaptation: Implications for a Changing Climate." *Journal of the Association of Environmental and Resource Economists* 7, no. 2 (2020): 209–42.

Easterling, J. "Family Turns Can of Worms into Booming Business." *Statesman Journal*, October 11, 1984, 15.

Edwards, C. A. "The Importance of Earthworms as Key Representatives of the Soil Fauna." In *Earthworm Ecology*, edited by C. A. Edwards, 2nd ed., 3–11. CRC Press, 2004.

Edwards, C. A., and J. R. Lofty. "Nitrogenous Fertilizers and Earthworm Populations in Agricultural Soils." *Soil Biology and Biochemistry* 14, no. 5 (1982): 515–21.

Egan, D. "Crawler Shock Hitting Local Anglers Right in the Worm Bucket." June 2014. http://www.cleveland.com/outdoors/index.ssf/2014/06/crawler _shock_hitting_local_an.html.

Ekers, M., and M. Farnan. "Planting the Nation: Tree Planting Art and the Endurance of Canadian Nationalism." *Space and Culture* 13, no. 1 (2010): 95–120.

Elson, D., and R. Pearson. "'Nimble Fingers Make Cheap Workers': An Analysis of Women's Employment in Third World Export Manufacturing." *Feminist Review* 7, no. 1 (1981): 87–107.

Engel-Di Mauro, S. *Ecology, Soils, and the Left: An Ecosocial Approach*. Springer, 2014.

FAO. *Soil Map of the World*. 1972. http://www.fao.org/soils-portal/soil-survey/soil -maps-and-databases/faounesco-soil-map of-the-world/en.

Federici, S. *Caliban and the Witch*. Autonomedia, 2004.

———. "Notes on Gender in Marx's *Capital*." 2017. https://ir.canterbury.ac.nz /bitstream/10092/14484/1/3%20Federici%20capital.pdf.

Fegely, T. "Canada's Drought Cuts Flow of Nightcrawlers." *The Morning Call*, July 5, 1988.

Felli, R. "On Climate Rent." *Historical Materialism* 22, no. 3–4 (2014): 251–80.

Ferguson, S. "Conditions Tough for Canada's Migrant Workers." *McLeans*, October 11, 2004. https://www.thecanadianencyclopedia.ca/en/article/conditions -tough-for-canadas-migrant-workers.

Ferraretto, L. F., and R. D. Shaver. "Meta-Analysis: Effect of Corn Silage Harvest Practices on Intake, Digestion, and Milk Production by Dairy Cows." *The Professional Animal Scientist* 28, no. 2 (2012): 141–49.

Ficke, A. D., C. A. Myrick, and L. J. Hansen. "Potential Impacts of Global Climate Change on Freshwater Fisheries." *Reviews in Fish Biology and Fisheries* 17, no. 4 (2007): 581–61.

The Field. "Worm Fishing for Fresh Trout." 1897, 595.

Fine, B. "On Marx's Theory of Agricultural Rent." *Economy and Society* 8, no. 3 (1979): 241–78.

———. "From Actor-Network Theory to Political Economy." *Capitalism Nature Socialism* 16, no. 4 (2005): 91–108.

Fischer, C., C. Roscher, B. Jensen, N. Eisenhauer, J. Baade, S. Attinger, S. Scheu, W. W. Weisser, J. Schumacher, and A. Hildebrandt. "How Do Earthworms, Soil Texture and Plant Composition Affect Infiltration along an Experimental Plant Diversity Gradient in Grassland?" *PloS One* 9, no. 6 (2014): e98987.

Fish and Wildlife Services. *1960 National Survey of Fishing and Hunting.* United States Department of Interior, 1960.

———. *1985 National Survey of Fishing and Hunting.* United States Department of Interior, 1985.

Foster, J. B. "Marx's Theory of Metabolic Rift: Classical Foundations for Environmental Sociology." *American Journal of Sociology* 105, no. 2 (1999): 366–405.

Foster, J. B., B. Clark, and R. York. *The Ecological Rift: Capitalism's War on the Earth.* NYU Press, 2011.

Foster, J., and A. Taylor. "In the Shadows: Exploring the Notion of 'Community' for Temporary Foreign Workers in a Boomtown." *Canadian Journal of Sociology* 38, no. 2 (2013): 167–90.

Fox, C. A., J. J. Miller, M. Joschko, C. F. Drury, and W. D. Reynolds. "Earthworm Population Dynamics as a Consequence of Long-Term and Recently Imposed Tillage in a Clay Loam Soil." *Canadian Journal of Soil Science* 97, no. 4 (2017): 561–79.

Frebman, P. "Rebate on Bait: Judge Decides What's Due for Dew Worm Loss." *Globe and Mail*, October 6, 1959, 19.

Freeman, R. B., and M. M. Kleiner. "The Last American Shoe Manufacturers: Decreasing Productivity and Increasing Profits in the Shift from Piece Rates to Continuous Flow Production." *Industrial Relations: A Journal of Economy and Society* 44, no. 2 (2005): 307–30.

Friends of Algonquin Park. "Special Fishing Regulations in Algonquin Park—Summary." N.d. https://www.algonquinpark.on.ca/visit/park_management/no-live-baitfish.php.

Fudge, J. "(Re) Conceptualising Unfree Labour: Local Labour Control Regimes and Constraints on Workers' Freedoms." *Global Labour Journal* 10, no. 2 (2019).

Fudge, J., and F. MacPhail. "The Temporary Foreign Worker Program in Canada: Low-Skilled Workers as An Extreme Form of Flexible Labour." *Comparative Labor Law and Policy Journal* 31 (2009): 101–39.

Furlong, M. A., D. R. Singleton, D. C. Coleman, and W. B. Whitman. "Molecular and Culture-Based Analyses of Prokaryotic Communities from an Agricultural Soil and the Burrows and Casts of the Earthworm Lumbricus Rubellus." *Applied and Environmental Microbiology* 68, no. 3 (2002): 1265–79.

Gabriel, C., and L. Macdonald. "After the International Organization for Migration: Recruitment of Guatemalan Temporary Agricultural Workers to Canada." *Journal of Ethnic and Migration Studies* 44, no. 10 (2018): 1706–24.

Gaupp-Berghausen, M., M. Hofer, B. Rewald, and J. G. Zaller. "Glyphosate-Based Herbicides Reduce the Activity and Reproduction of Earthworms and Lead to Increased Soil Nutrient Concentrations." *Scientific Reports* 5 (2015): 12886.

Giannaris, P. "Method and Medium for Coloring Live Bait Worms" (United States Patent No. US6240876B1), 2001. https://patents.google.com/patent/US6240876B1/en.

Gibson-Graham, J. K. *A Postcapitalist Politics.* University of Minnesota Press, 2006.

Gidwani, V. "The Cultural Logic of Work: Explaining Labour Deployment and Piece-Rate Contracts in Matar Taluka, Gujarat—Parts 1 and 2." *Journal of Development Studies* 38, no. 2 (2001): 57–108.

Globe and Mail (Toronto). "How to Earn $17.50: Get Light, Hunt Worms." April 9, 1955, 5.

———. "Worm Pickers Can Make Big Pay." May 5, 1956, 30.

———. Untitled. April 1, 1961, 23.

———. "Human Robins: Fast Worm-Picker Can Make Good Money, Live Bait Executive Says." June 29, 1964, 5.

———. "Ontario's Night Crawlers Outclass African Giants." September 9, 1971, 47.

———. "Beatings and Shotgun Fire Heating up Worm Wars." July 7, 1986.

Glover, J., and Y. Guerrier. "Women in Professional IT Jobs in the UK: Old Wine in New Bottles?" *Journal of Technology Management & Innovation* 5, no. 1 (2010): 85–94.

Goodman, D., and M. Redclift. "Capitalism, Petty Commodity Production and the Farm Enterprise." *Sociologia Ruralis* 25, no. 3-4 (1985): 231–47.

Goodman, D., B. Sorj, and J. Wilkinson. *From Farming to Biotechnology: A Theory of Agro-Industrial Development*. New York: Blackwell, 1987.

Goss, J. "Geography of Consumption I." *Progress in Human Geography* 28, no. 3 (2004): 369–80.

Government of Canada. "Guide 2201—Community Sponsors to Privately Sponsor Refugees." 2021. https://www.canada.ca/en/immigration-refugees-citizenship/services/application/application-forms-guides/guide-sponsor-refugee-community.html.

Government of Canada, S.C. "Tillage and Seeding Practices, Census of Agriculture, 2021." May 11, 2022. https://www150.statcan.gc.ca/t1/tbl1/en/tv.action?pid=3210036701.

Graff, O. "Darwin on Earthworms—The Contemporary Background and What the Critics Thought." In *Earthworm Ecology: From Darwin to Vermiculture*, edited by J. E. Satchell, 5–18. Springer Netherlands, 1983.

Grand Rapids Press. "Make Their Living by Digging Worms." August 13, 1919, 17.

Gregson, N. "And Now It's All Consumption?" *Progress in Human Geography* 19, no. 1 (1995): 135–41.

Guthman, J. *Wilted: Pathogens, Chemicals, and the Fragile Future of the Strawberry Industry*. University of California Press, 2019.

Hagey, R., U. Choudhry, S. Guruge, J. Turrittin, E. Collins, and R. Lee. "Immigrant Nurses' Experience of Racism." *Journal of Nursing Scholarship* 33, no. 4 (2001): 389–94.

Halford, F. *An Angler's Autobiography*. Vinton, 1903.

Hammon, H. "A Worm Farm Rewrites the Start-up Rules." *Financial Times*, 2010. https://www.ft.com/content/c881f070-4c1d-11df-a217-00144feab49a.

Hardt, M. "The Post-Operaist Approach to the Formal and Real Subsumption of Labor Under Capital." Seminar with Michael Hardt. *Praktyka Teoretyczna* 16 (2015): 167–82.

Hart, K. "Informal Income Opportunities and Urban Employment in Ghana." *The Journal of Modern African Studies* 11, no. 1 (1973): 61–89.

Harvey, D. "The Body as an Accumulation Strategy." *Environment and Planning D: Society and Space* 16, no. 4 (1998): 401–21.

———. *The Limits to Capital.* Verso, 2007.

Henderson, G. "Nature and Fictitious Capital: The Historical Geography of an Agrarian Question." In *Environment*, edited by Kay Anderson and Bruce Braun, 333–78. Routledge, 2017.

Herd, A. *The Fly.* Medlar Press, 2003.

Herod, A. "From a Geography of Labor to a Labor Geography: Labor's Spatial Fix and the Geography of Capitalism." *Antipode* 29, no. 1 (1997): 1–31.

———. "Labour Geography: Where Have We Been? Where Should We Go?" In *Missing Links in Labour Geography*, edited by NAMES, 15–28. Routledge, 2016.

Hodge, K. "Night Crawler Sellers Worm in on Fun, Profit." *The Times-News*, July 11, 1976, 53.

Hoefer, G., and K. H. Hartge. "Subsoil Compaction: Cause, Impact, Detection, and Prevention." In *Soil Engineering*, edited by A. P. Dedousis, and T. Bartzanas, 121–45. Springer, 2010.

Hoeffner, K., M. Santonja, D., Cluzeau, and C. Monard. "Epi-Anecic Rather Than Strict-Anecic Earthworms Enhance Soil Enzymatic Activities." *Soil Biology and Biochemistry* 132 (2019): 93–100.

Hoffmann, R. "Fishing for Sport in Medieval Europe: New Evidence." *Speculum* 60 (1985): 877–902.

———. "Economic Development and Aquatic Ecosystems in Medieval Europe." *The American Historical Review* 101, no. 3 (1996): 631–69.

———. "Trout and Fly, Work and Play in Medieval Europe." In *Backcasts: A Global History of Fly Fishing and Conservation*, edited by S. Snyder, B. Borgel, and E. Tobey, 27–45. University of Chicago Press, 2016.

Hoshmand, R. *Design of Experiments for Agriculture and the Natural Sciences.* Chapman and Hall/CRC, 2018.

Houle, R. "Results from the 2016 Census: Syrian Refugees Who Resettled in Canada in 2015 and 2016." Statistics Canada, 2019. https://www150.statcan.gc.ca/n1/pub/75-006-x/2019001/article/00001-eng.htm.

Hudson, R. "Cultural Political Economy Meets Global Production Networks: A Productive Meeting?" *Journal of Economic Geography* 8, no. 3 (2008): 421–40.

Hummel, R. L. *Hunting and Fishing for Sport: Commerce, Controversy, Popular Culture.* Popular Press, 1994.

Hunt, L. M., E. P. Fenichel, D. C. Fulton, R. Mendelsohn, J. W. Smith, T. D. Tunney, and J. E. Whitney. "Identifying Alternate Pathways for Climate Change to Impact Inland Recreational Fishers." *Fisheries* 41, no. 7 (2016): 362–72.

Hunt, L. M., and J. Moore. "The Potential Impacts of Climate Change on Recreational Fishing in Northern Ontario." Climate Change Research Report. Ontario Forest Research Institute, 2006.

Husar, J. "Fish-Bait Business Is Attractive, But U.S. Farmers Aren't Biting." *Chicago Tribune*, July 6, 1988, 45.

Irmler, U. "Changes in Earthworm Populations during Conversion from Conventional to Organic Farming." *Agriculture, Ecosystems & Environment* 135, no. 3 (2010): 194–98.

Jackson, K. "Best Way to Hook Kids Is with Worms." *Albany Democrat-Herald*, April 6, 1989, 18.

Jackson, O. R. "Worm Trap" (United States Patent No. US3543433A), 1970. https://patents.google.com/patent/US3543433A/en.

James, S. "When Nothing Else Works . . . Try a Worm." *Popular Mechanics* (March 1964), 117–18.

Jennings, J. *Alfalfa for Dairy Cattle*. University of Arkansas, 1996. https://www.uaex.uada.edu/publications/PDF/FSA-4000.pdf.

Jones, R., C. Travers, C. Rodgers, B. Lazar, E. English, J. Lipton, and J. Martinich. "Climate Change Impacts on Freshwater Recreational Fishing in the United States." *Mitigation and Adaptation Strategies for Global Change* 18, no. 6 (2013): 731–58.

Joyce, S. "Rediscovering the Cash Nexus, Again: Subsumption and the Labour–Capital Relation in Platform Work." *Capital & Class* 44, no. 4 (2020): 541–52.

Kalamazoo Gazette. "Sale of Angleworms to Fishermen Makes New Industry in City." July 9, 1912, 10.

———. "Kalamazoo Kiddie Makes Spending Money Selling Worms at Depots." August 8, 1914, 26.

Kallis, G., and E. Swyngedouw. "Do Bees Produce Value? A Conversation between an Ecological Economist and a Marxist Geographer." *Capitalism Nature Socialism* 29, no. 3 (2018): 36–50.

Kammenga, J. E., D. J. Spurgeon, C. Svendsen, and J. M. Weeks. "Explaining Density-Dependent Regulation in Earthworm Populations Using Life-History Analysis." *Oikos* 100, no. 1 (2003): 89–95.

Katz, C., and J. Monk. "When in the World Are Women?" 2008. https://atrium.lib.uoguelph.ca/bitstreams/c55caf79-d86d-4cc1-b1ed-bcb184ed67a0/download.

Kautsky, K. *The Agrarian Question: In Two Volumes* (Vol. 1). Zwan, 1988.

Keith, A.M., B. Boots, M.E. Stromberger, and O. Schmidt. "Consequences of Anecic Earthworm Removal over 18 Months for Earthworm Assemblages and Nutrient Cycling in a Grassland." *Pedobiologia* 66 (2018): 65–73.

Keller, R. P., A. N. Cox, C. Loon, D. M. Lodge, L. M. Herborg, and J. Rothlisberger. "From Bait Shops to the Forest Floor: Earthworm Use and Disposal by Anglers." *The American Midland Naturalist* 158, no. 2 (207): 321–28.

Kennedy, B. "Canada's Migrant Farm Worker Program Was Founded on 'Racist' Policies, New Lawsuit Alleges. And Today's Workers Are Still Paying the Price." *Toronto Star*, December 11, 2023. https://www.thestar.com/news/investigations/canada-s-migrant-farm-worker-program-was-founded-on-racist-policies-new-lawsuit-alleges-and/article_24d3281c-95II-11ee-8c5b-f3ce5c6d54cb.html.

Kirsch, S. "Cultural Geography I: Materialist Turns." *Progress in Human Geography* 37, no. 3 (2013): 433–41.

Kirsch, S., and D. Mitchell. "The Nature of Things: Dead Labor, Nonhuman Actors, and the Persistence of Marxism." *Antipode* 36, no. 4 (2004): 687–705.

Kirsch, V. "Migrant Workers Decry 'Slavery.'" *Guelph Mercury*, October 18, 2004, 3.

Kloppenburg, J. R. *First the Seed: The Political Economy of Plant Biotechnology*. University of Wisconsin Press, 2005.

Lamphier, G. "What Is Our Man Doing on Golf Course at 4 in the Morning?" *Wall Street Journal*, November 1, 1987, 11.

Lanphier, C. M. "Canada's Response to Refugees." *International Migration Review* 15, no. 1-2 (1981): 113–30.

Lavigne, Y. "'Black Market' Unearthed Theft Part of War for Worms." *The Globe and Mail*, August 15, 1979.

Lavkulich, L. M., and J. M. Arocena. "Luvisolic Soils of Canada: Genesis, Distribution, and Classification." *Canadian Journal of Soil Science* 91, no. 5 (2011): 781–806.

Le v. Canada, No. 1263 (Minister of Citizenship and Immigration). June 13, 2000.

LeBaron, G. "Unfree Labour beyond Binaries: Insecurity, Social Hierarchy and Labour Market Restructuring." *International Feminist Journal of Politics* 17, no. 1 (2015): 1–19.

Leff, E. *Green Production: Toward an Environmental Rationality*. Guilford Press, 1995.

Lerche, J. "The Unfree Labour Category and Unfree Labour Estimates: A Continuum within Low-End Labour Relations." *Manchester Papers in Political Economy*, CSPE (2011), 1–45.

Lindroth, C. *The Faunal Connections between Europe and North America*. Wiley, 1957.

Liu, Z., and X. Zou. "Exotic Earthworms Accelerate Plant Litter Decomposition in a Puerto Rican Pasture and a Wet Forest." *Ecological Applications* 12, no. 5 (2002): 1406–17.

Locker, A. "The Social History of Coarse Angling in England AD 1750–1950." *Anthropozoologica* 49, no. 1 (2014): 99–107.

Logt, R., C. Versteeg, P. Struyk, and N. Eekeren. "The Anecic Earthworm Lumbricus Terrestris Can Persist after Introduction into Permanent Grassland on Sandy Soil." *European Journal of Soil Biology* 119 (2023): 103536.

Lowe, C. N., and K. R. Butt. "Culture Techniques for Soil Dwelling Earthworms: A Review." *Pedobiologia* 49, no. 5 (2005): 401–13.

Lowe, C. N., K. R. Butt, and R. L. Sherman. "Current and Potential Benefits of Mass Earthworm Culture." In *Mass Production of Beneficial Organisms*, edited by J. A. Morales-Ramos, M. G. Rojas, and D. I.. Shapiro-Ilan, 683–709. Academic Press, 2014.

Lowerson, J. R. *Sport and the English Middle Classes, 1870–1914*. Manchester University Press, 1993.

MacDonald, J. L. *Aspects of Iu-Mien Refugee Identity.* Taylor & Francis, 1997.

Magalhaes, L., C. Carrasco, and D. Gastaldo. "Undocumented Migrants in Canada: A Scope Literature Review on Health, Access to Services, and Working Conditions." *Journal of Immigrant and Minority Health* 12, no. 1 (2010): 132.

Maldonado, M. M. "'It Is Their Nature to Do Menial Labour': The Racialization of 'Latino/a Workers' by Agricultural Employers." In *Latino Identity in Contemporary America*, edited by M. Bulmer, and J. Solomos, 93–112. Routledge, 2012.

Mann, S. A., and J. M. Dickinson. "Obstacles to the Development of a Capitalist Agriculture." *The Journal of Peasant Studies* 5, no. 4 (1978): 466–81.

Martin, P. "Migrant Workers in Commercial Agriculture." International Labor Office, 2016. https://www.ilo.org/wcmsp5/groups/public/---ed_protect/---protrav/---migrant/documents/publication/wcms_538710.pdf.

Marx, K. *Capital: Volume I.* Penguin Classics, 1990.

———. *Capital: Volume III.* Penguin UK, 1992.

Marysville Journal-Tribune. "Return of Walleyes Brings Worm Wars." *Marysville Journal-Tribune*, August 6, 1987, 2.

Mas, S. "Do Government-Assisted Refugees Receive More Money for Food than Canadians on Welfare?" *CBC News*, 2016. https://www.cbc.ca/news/politics/do-government-assisted-refugees-receive-more-money-for-food-than-canadians-on-welfare-1.3230503.

Mazzucato, M. *The Value of Everything: Making and Taking in the Global Economy* (Illustrated edition). PublicAffairs, 2018.

McCarthy, J. "Political Ecology/Economy." In *The Wiley-Blackwell Companion to Economic Geography*, edited by Trevor J. Barnes, Jamie Peck, and Eric Sheppard, 612–25. Wiley-Blackwell, 2012.

McDowell, L., A. Batnitzky, and S. Dyer, S. "Division, Segmentation, and Interpellation: The Embodied Labors of Migrant Workers in a Greater London Hotel." *Economic Geography* 83, no. 1 (2007): 1–25.

McGrath, S. "Unfree Labor." In *International Encyclopedia of Geography*, edited by D. Richardson, N. Castree, M. F. Goodchild, A. Kobayashi, W. Liu, and R. A. Marston, 1–8. John Wiley & Sons, 2017. https://doi.org/10.1002/9781118786352.wbieg0703.

McLean, N. *A River Runs Through It and Other Stories.* University of Chicago Press, 2016.

Melamed, J. "Racial Capitalism." *Critical Ethnic Studies* 1, no. 1 (2015): 76.

Mezzadri, A. "A Value Theory of Inclusion: Informal Labour, the Homeworker, and the Social Reproduction of Value." *Antipode* 53, no. 4 (2021): 1186–1205.

Mies, M. *Patriarchy and Accumulation on a World Scale: Women in the International Division of Labour.* Bloomsbury, 2014.

Milcu, A., S. Partsch, C. Scherber, W. W. Weisser, and S. Scheu. "Earthworms and Legumes Control Litter Decomposition in a Plant Diversity Gradient." *Ecology* 89, no. 7 (2008): 1872–82.

Miller, J. "The Golf Course Worm Is an Educated Worm." *Calgary Herald*, November 27, 1971, 71–72.

Miller, J.J., M.L. Owen, C.F. Drury, and D.S. Chanasyk. "Short-Term Legacy Effects of Feedlot Manure Amendments on Earthworm Abundance in a Clay Loam Soil." *Canadian Journal of Soil Science* 99, no. 4 (2019): 447–57.

Mills, M.B. "From Nimble Fingers to Raised Fists: Women and Labor Activism in Globalizing Thailand." *Signs: Journal of Women in Culture and Society* 31, no. 1 (2005): 117–44.

Minneapolis Tribune. "Shortage of Pickers Hurts Worm Farm." July 19, 1970, 13B.

Mishra, P. "Urbanisation Through Brick Kilns: The Interrelationship Between Appropriation of Nature and Labour Regimes." *Urbanisation* 5, no. 1 (2020): 17–36.

Mitchell, D. "Dead Labor and the Political Economy of Landscape—California Living, California Dying." *Handbook of Cultural Geography* (2003): 233–48.

———. "Labor's Geography: Capital, Violence, Guest Workers and the Post-World War II Landscape." *Antipode* 43, no. 2 (2011): 563–95.

———. *They Saved the Crops: Labor, Landscape, and the Struggle over Industrial Farming in Bracero-Era California.* University of Georgia Press, 2012.

Mitchell, T. *Rule of Experts: Egypt, Techno-Politics, Modernity.* University of California Press, 2002.

Monteiro, L. "Mexican Couple Fear Death If They Go Home Again; Pair Who Complained about Working Conditions at Worm-Picking Operation in Guelph Seeking Refugee Status." *Guelph Mercury*, April 12, 2006, 4.

Montgomery, D. *Dirt: The Erosion of Civilizations.* University of California Press, 2007.

———. *Growing a Revolution: Bringing Our Soil Back to Life.* W.W. Norton & Company, 2017.

Montreal Star. "Worm Picking Big Business Now." June 16, 1959, 55.

Moore, J.W. "Ecological Crises and the Agrarian Question in World-Historical Perspective." *Monthly Review* 60, no. 6 (2008): 54–63.

———. *Capitalism in the Web of Life: Ecology and the Accumulation of Capital.* Verso Books, 2015.

Mt. Pleasant, J. "A New Paradigm for Pre-Columbian Agriculture in North America." *Early American Studies* (2015): 374–412.

———. "The Paradox of Plows and Productivity: An Agronomic Comparison of Cereal Grain Production under Iroquois Hoe Culture and European Plow Culture in the Seventeenth and Eighteenth Centuries." *Agricultural History* 85, no. 4 (2011): 460–92.

Mt. Pleasant, J., and R.F. Burt. "Estimating Productivity of Traditional Iroquoian Cropping Systems from Field Experiments and Historical Literature." *Journal of Ethnobiology* 30, no. 1 (2010): 52–79.

Müller, M. "Assemblages and Actor-Networks: Rethinking Socio-material Power, Politics and Space." *Geography Compass* 9, no. 1 (2015): 27–41.

Murray, P. "The Social and Material Transformation of Production by Capital: Formal and Real Subsumption in Capital." In *The Constitution of Capital*, edited by Riccardo Bellofiore and Nicola Taylor, 247–73. Palgrave Macmillan, 2004.

Murray, R. "Value and Theory of Rent: Part One." *Capital & Class* 1, no. 3 (1977): 100–22.

Musshauser, D., A. Bader, B. Wildt, and M. Hochleitner. "The Impact of Sociodemographic Factors vs. Gender Roles on Female Hospital Workers' Health: Do We Need to Shift Emphasis?" *Journal of Occupational Health* 48, no. 5 (2006): 383–91.

New Haven Evening Register. "Bring Out the Worms." April 18, 1891, 1.

New York Times. "Bait Fishing Popular." August 9, 1891, 2.

———. "Shave Pays for Fish Worms." June 15, 1912, 24.

———. "President Bait Wins Jersey Fishing Bout." July 8, 1927, 1.

———. "Barber Invents Machine to Vend His Fish Bait." June 23, 1936, 25.

Newman, P. "The Hungarians." *Macleans,* February 16, 1957, 12.

News-Journal (Wilington, Ohio). "Classified Ad." June 22, 1954, 12.

Nuutinen, V., and K. R. Butt. "Worms from the Cold: Lumbricid Life Stages in Boreal Clay during Frost." *Soil Biology and Biochemistry* 41, no. 7 (2009): 1580–82.

Nuutinen, V., M. Hagner, H. Jalli, L. Jauhiainen, S. Rämö, I. Sarikka, and J. Uusi-Kämppä. "Glyphosate Spraying and Earthworm Lumbricus Terrestris L. Activity: Evaluating Short-Term Impact in a Glasshouse Experiment Simulating Cereal Post-Harvest." *European Journal of Soil Biology* 96 (2020): 103148.

O'Connor, J. "Capitalism, Nature, Socialism: A Theoretical Introduction." *Capitalism, Nature, Socialism* 1, no. 1 (1988): 11–38.

OMAFRA. "Number of Milk Producers by County." 2021. http://www.omafra .gov.on.ca/english/stats/dairy/index.html.

Onrust, J., and T. Piersma. "How Dairy Farmers Manage the Interactions between Organic Fertilizers and Earthworm Ecotypes and Their Predators." *Agriculture, Ecosystems & Environment* 273 (2019): 80–85.

Orzeck, R. "What Does Not Kill You: Historical Materialism and the Body." *Environment and Planning D: Society and Space* 25, no. 3 (2007): 496–514.

Ouma, S., L. Johnson, and P. Bigger. "Rethinking the Financialization of 'Nature.'" *Environment and Planning A: Economy and Space* 50, no. 3 (2018): 500–11.

Papp-Zubrits, S. "The Forgotten Generation: Canada's Hungarian Refugees of 1956." *Oral History Forum d'histoire Orale* 4, no. 2 (1980): 29–34.

Parker, R. "Southwestern Ontario Land Values: 2020 Edition." 2020. http://www .valcoconsultants.com/wp-content/uploads/2021/02/Land-Values-Study-2020 .pdf.

Pfeifer, M. E. "Community, Adaptation, and the Vietnamese in Toronto." PhD diss., University of Toronto, 1999.

Pfiffner, L., and H. Luka. "Earthworm Populations in Two Low-Input Cereal Farming Systems." *Applied Soil Ecology* 37, no. 3 (2007): 184–91.

Pfiffner, L., and P. Mäder. "Effects of Biodynamic, Organic and Conventional Production Systems on Earthworm Populations." *Biological Agriculture & Horticulture* 15, no. 1-4 (1997): 2–10.

Pitcher, T. J., and C. E. Hollingworth. "Fishing for Fun: Where's the Catch?" In *Recreational Fisheries: Ecological, Economic and Social Evaluation,* edited by T. J. Pitcher and C. E. Hollingworth, 1–16. Blackwell Science Ltd., 2002.

Plain Dealer (Cleveland, Ohio). "Gathering Worms for Bait. Some Fishermen Have Novel Methods of Procuring A Supply." June 7, 1896, 27.

Portes, A., and W. Haller. "The Informal Economy." In *The Handbook of Economic Sociology*, 2nd ed., edited by N.J. Smelser and R. Swedberg, 403–28. Princeton University Press, 2010.

Portes, A., and S. Sassen-Koob. "Making It Underground: Comparative Material on the Informal Sector in Western Market Economies." *American Journal of Sociology* 93, no. 1 (1987): 30–61.

Preibisch, K. "Pick-Your-Own Labor: Migrant Workers and Flexibility in Canadian Agriculture 1." *International Migration Review* 44, no. 2 (2010): 404–41.

Proby, J. "Pickers Stealing His Worms, Farmer Says." *Toronto Star*, June 6, 1989, 7.

Prudham, W. S. *Knock on Wood: Nature as Commodity in Douglas-Fir Country*. Routledge, 2012.

Puzic, S. "Fact Check: Do Refugees Get More Financial Help Than Canadian Pensioners?" *CTV News*, 2015. https://www.ctvnews.ca/politics/fact-check-do-refugees -get-more-financial-help-than-canadian-pensioners-1.2670735.

R V Tran, No. 5747 (Ontario Court of Justice). 2020.

Raines, H. "Fishing with Presidents." *New York Times*, September 5, 1993, 18.

Reid-Musson, E. "Grown Close to Home™: Migrant Farmworker (Im) Mobilities and Unfreedom on Canadian Family Farms." *Annals of the American Association of Geographers* 107, no. 3 (2017): 716–30.

Reimer, G., and J.-P. Chartrand. *A Historical Profile of the James Bay Area's Mixed European-Indian or Mixed European-Inuit Community*. Praxis Research Associates, 2005.

Reynolds, J. *The Earthworms (Lumbricidae and Sparganophilidae) of Ontario*. Royal Ontario Museum, 1977.

———. "The Distribution of Earthworms (Annelida, Oligochaeta) in North America." In *Advances in Ecology and Environmental Sciences*, edited by P.C. Mishra, N. Behera, B.K. Senapati, and B.C. Guru, 133–53. Ashish Publishing House, 1995.

Ribot, J.C., and N.L. Peluso. "A Theory of Access." *Rural Sociology* 68, no. 2 (2003): 153–81.

Rogaly, B. "Migrant Workers in the ILO's Global Alliance Against Forced Labour Report: A Critical Appraisal." *Third World Quarterly* 29, no. 7 (2008): 1431–47.

Rupp, V. "Hunting and Fishing." *Bend Bulletin*, March 5, 1958, 2.

Sahrhage, D., and J. Lundbeck. *A History of Fishing*. Springer Science & Business Media, 1992.

Schneider, M., and P. McMichael. "Deepening, and Repairing, the Metabolic Rift." *The Journal of Peasant Studies* 37, no. 3 (2010): 461–84.

Shaer, L. "Migrant Worker Programs a Win-Win for Workers and Farmers." *National Post*, November 16, 2021. https://nationalpost.com/sponsored/news-sponsored /migrant-worker-programs-a-win-win-for-workers-and-farmers.

Sharpley, A., R. McDowell, B. Moyer, and R. Littlejohn. "Land Application of Manure Can Influence Earthworm Activity and Soil Phosphorus Distribution." *Communications in Soil Science and Plant Analysis* 42, no. 2 (2011): 194–207.

Sherman, R. *The Worm Farmer's Handbook: Mid- to Large-Scale Vermicomposting for Farms, Businesses, Municipalities, Schools, and Institutions.* Chelsea Green Publishing, 2018.

Siewert, H. F., and J. B. Cave. "Survival of Released Bluegill, Lepomis Macrochirus, Caught on Artificial Flies, Worms, and Spinner Lures." *Journal of Freshwater Ecology* 5, no. 4 (1990): 407–11.

Silva-Costa, A., R. H. Griep, F. M. Fischer, and L. Rotenberg. "Need for Recovery from Work and Sleep-Related Complaints among Nursing Professionals." *Work* 41 (Supplement 1) (2012): 3726–31.

Simner, M. L. *How Middlesex County Was Settled with Farmers, Artisans, and Capitalists: An Account of the Canada Land Company in Promoting Emigration from the British Isles in the 1830s through the 1850s.* London and Middlesex Heritage Museum, 2010.

Smith, C. "The Double Indeterminacy of Labour Power: Labour Effort and Labour Mobility." *Work, Employment and Society* 20, no. 2 (2006): 389–402.

Smith, N. "Nature as Accumulation Strategy." *Socialist Register* (2007): 43.

———. *Uneven Development: Nature, Capital, and the Production of Space.* University of Georgia Press, 2008.

Star Tribune (Minneapolis, Minnesota). "Business Goes as the Worms Turn." February 21, 1988, 11B.

Statesman Journal (Salem, Oregon). "Worm Hunters Ruin Flowers." June 4, 1959, 3.

Statistics Canada. "Minimum Wage in Canada since 1975." 2018. https://www150.statcan.gc.ca/n1/pub/11-630-x/11-630-x2015006-eng.htm.

Steckley, J. "Nightcrawler Commodities: A Brief History on the Commodification of the Humble Dew Worm." *Environment and Planning E: Nature and Space* 5, no. 3 (2022): 1361–82.

——— "'Completely Free': How a Subsumption of Labour and Nature Framework Explains the Surprising Expressions of 'Freedom' by Immigrant Worm pickers in Ontario." *Antipode*, 2024.

Stellin, F., F. Gavinelli, P. Stevanato, G. Concheri, A. Squartini, and M. G. Paoletti. "Effects of Different Concentrations of Glyphosate (Roundup 360®) on Earthworms (Octodrilus Complanatus, Lumbricus Terrestris and Aporrectodea Caliginosa) in Vineyards in the North-East of Italy." *Applied Soil Ecology* 123 (2018): 802–8.

Stewart, A. *The Earth Moved: On the Remarkable Achievements of Earthworms.* Algonquin Books, 2005.

Strauss, K. "Labour Geography III: Precarity, Racial Capitalisms and Infrastructure." *Progress in Human Geography* 44, no. 6 (2020): 1212–24.

Strauss, K., and S. McGrath. "Temporary Migration, Precarious Employment and Unfree Labour Relations: Exploring the 'Continuum of Exploitation' in Canada's Temporary Foreign Worker Program." *Geoforum* 78 (2017): 199–208.

Szymkowiak, M. "Genderizing Fisheries: Assessing over Thirty Years of Women's Participation in Alaska Fisheries." *Marine Policy* 115 (2020): 103846.

Taekema, D. "Worm Picking a Slippery Industry in Toronto." *The Toronto Star*, October 25, 2015. https://www.thestar.com/news/gta/2015/10/25/worm-picking -a-slippery-industry-in-toronto.html.

Taylor, M., and S. Rioux. *Global Labour Studies*. John Wiley & Sons, 2017.

Telford, R. "How the West Was Won: Land Transactions between the Anishinabe, the Huron and the Crown in Southwestern Ontario." *Algonquian Papers-Archive*, 1998, 29. https://ojs.library.carleton.ca/index.php/ALGQP/article/download /491/393.

Thammavongsa, Souvankham. *How to Pronounce Knife: Stories*. McClelland & Stewart, 2020.

The Times-Picayune (Louisiana, New Orleans). "Recent Discovery Ends Bait Digging. Lucedale Man Finds Method of Coaxing Angle Worms out of the Ground." June 14, 1914, 1.

———. "'Bait' Sales Make Monroe Lads Rich." July 26, 1920, 4.

Thompson, E. P. *Whigs and Hunters: The Origin of the Black Acts*. Pantheon Books, 1975.

Times-News (Twin Falls, Idaho). "Breathless Worms Make Student Money." August 7, 1994, 31.

Tomba, M. "On the Capitalist and Emancipatory Use of Asynchronies in Formal Subsumption." *Review (Fernand Braudel Center)* 38, no. 4 (2015): 287–306.

Tomlin, A.D. "The Earthworm Bait Market in North America." In *Earthworm Ecology: From Darwin to Vermiculture*, edited by J. Satchell, 331–38. Chapman and Hall, 1983.

Tomlin, A.D., and C.A. Fox. "Earthworms and Agricultural Systems: Status of Knowledge and Research in Canada." *Canadian Journal of Soil Science* 83 (Special Issue) (2003): 265–78.

Tomlin, A.D., and J.J. Miller. "Impact of Ring-Billed Gull (*Larus Delawarensis Ord.*) Foraging on Earthworm Populations of Southwestern Ontario Agricultural Soils." *Agriculture, Ecosystems & Environment* 20, no. 3 (1988): 165–73.

Tomlin, A., and F. Gore. "Effects of Six Insecticides and a Fungicide on the Numbers and Biomass of Earthworms in Pasture." *Bulletin of Environmental Contamination and Toxicology* 12, no. 4 (1974): 487–92.

Trenton Evening Times. "Why Dig Fish Worms." August 28, 1922, 2.

Troper, H. "History of Immigration to Toronto Since the Second World War: From Toronto 'the Good' to Toronto 'the World in a City.'" CERIS Working Paper 12 (2000), 1–37.

Tsing, A. "Free in the Forest: Popular Neoliberalism and the Aftermath of War in the US Pacific Northwest." In *Rhetorics of Insecurity: Belonging and Violence in the Neoliberal Era*, edited by Zeynep Gambetti and Marcial Godoy-Anativia, 20–39. New York University Press and Social Science Research Council, 2013.

———. *The Mushroom at the End of the World: On the Possibility of Life in Capitalist Ruins*. Princeton University Press, 2015.

UNHCR. *Refugees in Canada*. N.d. https://www.unhcr.ca/in-canada/refugees- in-canada.

———. "Seven Decades of Refugee Protection in Canada: 1950–2020." 2020. https://www.unhcr.ca/wp-content/uploads/2020/12/Seven-Decades-of-Refugee-Protection-In-Canada-14-December-2020.pdf.

United States Court of Appeals. Number 15-5175. https://www.opn.ca6.uscourts.gov/opinions.pdf/15a0242p-06.pdf.

USA Trade Online. United States Census Bureau, 2022.

Van Groenigen, J.W., I.M. Lubbers, H.M. Vos, G.G. Brown, G.B. De Deyn, and K.J. Van Groenigen. "Earthworms Increase Plant Production: A Meta-Analysis." *Scientific Reports* 4, no. 1 (2014): 1–7.

Velásquez, D., and J. Ayala. "Production of Nature and Labour Agency: How the Subsumption of Nature Affects Trade Union Action in the Fishery and Aquaculture Sectors in Aysén, Chile." *Environment and Planning E: Nature and Space* 7, no. 2 (2024): 720–41.

Vercellone, C. "From Formal Subsumption to General Intellect: Elements for a Marxist Reading of the Thesis of Cognitive Capitalism." *Historical Materialism* 15, no. 1 (2007): 13–36.

Vincent, L.A., X. Zhang, É. Mekis, H. Wan, and E.J. Bush. "Changes in Canada's Climate: Trends in Indices Based on Daily Temperature and Precipitation Data." *Atmosphere-Ocean* 56, no. 5 (2018): 332–49.

Volgenau, G. "Worm Wars." *Detroit Free Press*, July 17, 1986, 74, 1.

Vosko, L.F., and R. Casey. "Enforcing Employment Standards for Temporary Migrant Agricultural Workers in Ontario, Canada: Exposing Underexplored Layers of Vulnerability." *International Journal of Comparative Labour Law and Industrial Relations* 35, no. 2 (2019).

Walker, R. "Value and Nature: From Value Theory to the Fate of the Earth." *Human Geography* 9, no. 1 (2016): 1–15.

———. "Value and Nature: Rethinking Capitalist Exploitation and Expansion." *Capitalism Nature Socialism* 28, no. 1 (2017): 53–61.

Washabaugh, W., and C. Washabaugh. *Deep Trout: Angling in Popular Culture.* Routledge, 2000.

Washington Post. "Golf Course Gleaners Stalk Worms, Not Ball." July 7, 1956, A19.

Watts, N., and I.R. Scales. "Seeds, Agricultural Systems and Socio-Natures: Towards an Actor–Network Theory Informed Political Ecology of Agriculture." *Geography Compass* 9, no. 5 (2015): 225–36.

Webster, A. "Willing Worms Wait for Warm, Windless Weather." *Globe and Mail*, June 13, 1966, 5.

Whalen, J.K., R.W. Parmelee, and C.A. Edwards. "Population Dynamics of Earthworm Communities in Corn Agroecosystems Receiving Organic or Inorganic Fertilizer Amendments." *Biology and Fertility of Soils* 27, no. 4 (1998): 400–7.

White, R. *The Organic Machine: The Remaking of the Columbia River.* Hill and Wang, 2011.

Williams, M.J. "How Are Fisheries and Aquaculture Institutions Considering Gender Issues?" *Asian Fisheries Science* (2016).

Wilson, T. D. "Precarization, Informalization, and Marx." *Review of Radical Political Economics* 52, no. 3 (2020): 470–86.

Windsor Star. "Worms Rejoice." June 22, 1966, 20.

Wolfthal, D. "Shedding Some Light on the Night Crawler, Still a Fine Bait." *New York Times*, April 27, 1980, 474.

Wood, E. M. *The Origin of Capitalism: A Longer View.* Verso, 2002.

Wright, M. *Disposable Women and Other Myths of Global Capitalism.* Routledge, 2013.

INDEX

absolute surplus value. *See* surplus value

Actor Network Theory (ANT), 26, 191

agency: of capital, 26–27, 191, 200; and ecology, 160–161, 193, 199; of labor, 26–27, 145, 158, 160–161, 165, 169–170, 193; of nonhuman actors, 14, 26–27

agrarian question, 13

alfalfa: within crop rotations, 73–74, 77, 101; dried and fermented forms of, 95–96; importance in dairy farming of, 72, 95–97; as a market commodity, 92–94, 96–97; relationship to worm contracts, 85–88, 93. 97, 102; relationship to worm populations, 28, 55–56, 73, 79, 82, 179, 197

anglers: demography of, 177–178, 188, 200; hierarchy of, 31, 39–41, 43; number of recreational, 42, 182–183; preference for nightcrawlers, 6, 8–9, 31, 43, 46, 49, 52, 56, 176; selection of bait, 7, 41, 44, 56, 185–186. *See also* fishermen

angling: associations for, 124; class dynamics of, 40, 42; conventions of, 31, 34; definition of, 39; for food, 39–40; popularity of, 40; for recreation, 40–43, 182, 187–188; types of bait for, 43. *See also* fishing

appropriation: as a concept in critical agrarian studies, 13, 78; of energy/ work, 16; of surplus value, 12, 22, 29, 194–195

assemblage: benefits and drawbacks of the concept, 27–29, 198; definition of,

26–27; in relation to bait industry, 32–33, 56, 74, 76, 101, 178, 188, 191–193, 199–200; in relation to capital, 28–30, 134, 192, 200

autonomy: in the labor process, 138, 158, 161, 169; in relation to subsumption, 138, 161–162, 172–173; in worm picking, 22, 28, 32, 157, 160, 168, 172–173

bait: artificial, 121, 185–187; characteristics of, 43–44, 58–61, 121–122; as commodities, 46–47, 58, 61, 105, 121–123, 192; companies, 50–52; demand for, 39, 56, 59, 183; digging for, 45–46; effectiveness of, 43, 184; hierarchy of, 40–43; history of, 31; live, 41, 43, 122, 124, 183–184, 201n9; nightcrawlers as, 3, 10, 42; most popular, 7, 9, 10, 43, 185; price of, 6, 8, 125, 182; retailers/sellers of, 4, 8, 30, 46, 52, 107, 120, 124, 177; regulation of, 183–184, 200; types of, 25, 55, 56–60

bait worm industry. *See* nightcrawler industry

bait worms. See *Lumbricus terrestris*; nightcrawlers

ballast, 26, 28, 61

barriers to entry, 23, 26, 32, 105–107, 118–119

bees, 16

Berkley (business), 186

biker gangs, 106, 131

Boyd, William, 18, 19, 20

Canada: and colonization, 61–64; dairy industry in, 76–77, 94, 98, 102, 180; earthworms in, 61–63; history of nightcrawler industry in, 47–50; immigration to, 50–51, 108, 120, 149, 151, 175, 190; import of nightcrawlers from, 45, 52, 185, 201n9; refugees in, 8, 12, 151–154, 156; residency status in, 165–168, 175, 199; soils in, 49; tax laws in, 128; Temporary Foreign Worker Program in, 169; weather in, 187

Canada Borders and Customs Agency (CBCA), 167–168

Canada Revenue Agency (CRA), 5, 126–129, 131, 167–168, 214n18

Canadian Broadcasting Company, 184

capital: accumulation in the bait industry, 15, 28, 52, 102, 195, 200; accumulation through nature, 12–17, 19, 22–25, 79, 200; accumulation through social differentiation, 147–148; agency of, 26–28, 191, 200; of coolers, 15; experimental nature of, 33, 74, 192–193; investment of, 24–25, 49, 106, 110, 118–120, 129; and labor geography, 160–161; necessary for red wiggler production, 56–57, 59–60; as part of assemblage, 26–30, 33, 76, 178; relation to rent, 76–79; struggle with labor, 15, 151, 165; and subsumption, 17–19, 22–24, 78–79, 134, 136, 161–162, 190, 199–200; understandings of, 27

capitalism: in agriculture, 12–14; concept of subsumption within, 12, 191, 194; dependence on nature and labor, 16–17, 24, 194; and the environment, 13–14, 16, 157–158; freedom within, 159–161; nightcrawler industry within, 200; racial, 147; rent within contemporary, 77–79

Carter, Hugh, 56–57, 59

Carter, Jimmy, 26, 56

cash: advances in, 165; bags of, 1, 5, 6, 90, 98, 125, 180; farmer income in, 90–91; payment to farmers in, 5, 7, 69–70, 77, 90–91; transactions in the nightcrawler industry, 4–5, 125–126; in underground economy, 105, 126–127, 130–131, 168, 190, 198; value of the worms, 81; worm-

picker income in, 156–157, 165–168, 165–167

cash crops: challenges in production of, 75; farmers of, 98, 180–181, 197; income from, 91–93; and worm populations, 67; worms as, 7, 96, 180–181, 196

catfish, 39, 431

cattle feed. *See* alfalfa

checks: benefits of being paid by, 112–113, 166–167; paid to farmers, 77, 84, 88, 127; paid to warehouse workers, 112–113, 131; paid to worm pickers, 165–166; relationship to immigration status, 167; in an underground economy, 130–131

China, 133

climate change, 26, 32, 176, 178, 187–188, 195, 200

colonization, 55, 61–64

commodification: capitalist forms of, 33–38; definition of, 37–38; historical process of, 37, 52; of *Lumbricus terrestris*, 46–52, 57; of nature, 195; open-ended process of, 39, 52; stories of, 37, 53

commodity production: in agriculture, 13–14; motivated by exchange value, 38; relationship to nature, 16–17, 160; and the subsumption of nature, 18–19, 22–23

constrained agency, 160, 166, 172, 193

consumption within capitalism, 37, 102, 195, 205n2

coolers: ability to profit, 55, 162; competition between, 31; under COVID lockdowns, 177–178, 182; description of, 8, 123–125; estimate on total number of, 201n9; export requirements of, 184–185; how to become, 107–110; investment decisions by, 110, 113–114, 118–120; leases with farmers, 73, 84–88, 92, 114, 165, 179–182; leases with golf courses, 68; and money laundering, 129–132; payment schemes, 144–145, 163, 165; perspectives on climate change, 187; perspectives on weather, 116–117, 164; perspectives on worm pickers, 147–149, 178, 189–190; relationship with customers, 120, 182–183; and tax evasion, 126–129; use of technology, 117–119; use of temporary foreign laborers, 159, 166–

impact on soil, 14, 49, 55, 64–65; impact of tillage on, 66–69, 72–73; introduction into North America, 61–64; populations in soil, 55, 68; and rent, 23, 79, 180, 190; research on, 81–82, 196–197; transition to a commodity, 32, 37–38, 52–53. *See also* earthworms; nightcrawlers

mafia, 130
manure: and nutrient management plans, 94, 97; as part of assemblage, 25, 28, 74, 180–181; and worm-picking contracts, 84, 86, 89–90, 97, 101, 114; and worm populations, 7, 68, 72–73, 76, 102, 162
Marx, Karl: on capitalist value, 15, 18, 200; on commodities, 37; on the Factory Acts, 106; on freedom, 158–159, 161; on nature, 13, 15–16; on piece rates, 144–145; on rent, 77; on subsumption, 15–19, 105, 136, 203n25
materiality: of nature, 12, 14; of nightcrawler production, 32, 166, 173, 193; of production, 22–24; relation to labor process, 158–160, 168
Mazzucato, Mariana, 77
McLean, Norman, 41
Mennonites, 83, 212n16
Mills, Mary Beth, 150
money laundering, 9, 32, 129–132
Moore, Jason, 16
Mill, Jon Stuart, 203n28
Mitchell, Don, 169
Mitchell, Timothy, 27
Montgomery, David, 67
Mushroom at the End of the World, The (A. Tsing), 29

National Survey of Fishing and Hunting, 42
nature: and assemblages, 26–30, 192, 198; capitalist control over, 19–20, 23, 52, 154, 193; and capitalist value, 15–20, 33; as constitutive of social relations, 14–15, 28–29, 191–194; and land rent, 78–79; materiality of, 12–14, 22–24, 158, 193; pricing of, 16; production of, 19–20, 25; productivity of, 19–20; relationship to

capitalism, 12–20, 22–26; subsumption of, 17–20, 22–25, 30, 78, 106, 158–162, 193–195
Netherlands, 59, 185, 197
New England, 61, 188
New York Times, 41, 43, 46, 47
newspapers: documenting the early industry, 31, 35, 45–47, 148, 171–172; interest in the bait industry, 4, 126, 184, 199
Niagara region, 10, 11
nightcrawler capitalists: access to labor, 50, 52, 151–154, 199; access to land, 55, 68–69, 76, 101–103, 117–118, 179–180; description of, 8; frustrations with worm commodity, 12, 22–23, 25–26, 74, 106, 193; in history, 47; in the informal economy, 32, 125–131; investments by, 11, 24, 106–114, 119; in the nightcrawler assemblage, 28, 56, 73, 178, 191–192; un/control over labor, 32, 136, 144–145, 158, 162, 164–165, 169–175; at the worm exchange, 36–37. *See also* coolers; crew chiefs
nightcrawler industry: as an assemblage, 28–30, 33, 74, 178–179, 188, 192–193, 200; as a case study of capitalist development, 190–191; CRA's knowledge of, 5, 126–129, 131, 168, 214n18; dependence on other industries, 56, 162; description of, 4–9; economic value of, 120, 195, 201n9; emergence of rent in, 76–77, 102; emergent properties of, 32–33, 36, 49, 190; entry into, 35, 106–108; human and nonhuman actors in, 6–9; labor in, 50–51, 151–155; money laundering in, 129–132; peculiarities of, 18, 23, 39, 89, 173, 190–191; reliance on dairy farms/farmers, 76; role of *L. terrestris* in shaping, 12, 15, 190–192; tax evasion in, 127–129; technology in, 11, 23, 119, 132, 136–142; temporary foreign workers in, 169–175; threats to, 32, 178–188; transportation networks of, 114; underground nature of, 104–105, 126–132, 168, 199; and the Vietnam War, 151–153
nightcrawlers (as commodity): bait characteristics of, 43–44; as a cash crop, 7, 73, 76; commodification of, 38–39, 102; COVID demand for, 177, 188; cultivation

89–90; relationship to subsumption, 23–24, 78–79, 118, 181; as social-ecological relation, 78–79, 102, 195

rentiers: farmers as accidental, 31, 76–77, 79, 97, 102–103, 179, 194; parasitic conception of, 77–78, 198

residency status: impact on conceptions of freedom, 165–168, 172–173, 175, 199; of pickers, 165–173

Reynolds, John, 61, 146

"River Runs Through It, A" (N. McLean), 41–42

Roundup, 65–66, 84

Royal Canadian Mounted Police, 5

salmon, 24, 40–41

Schurman, Rachel, 18

Silent Spring (R. Carson), 65

Silver Bait (company), 59–60

Smith, Neil, 19

Social Insurance Numbers (SINs), 126

socially necessary labor time, 15–17, 31, 39, 47, 194

soil: in bait assemblage, 6, 26, 32, 101, 188, 200; capitalism's impact on, 13, 22: compaction of, 67, 101, 180; as earthworm's habitat, 57–61, 78; earthworm's impact on, 7, 12, 14, 28, 64–65, 73, 80, 99–102, 184–185; fertility of, 77, 79, 98, 101–102, 180; as instrument of nightcrawler production, 30, 118, 134, 192; precolonial, 61–63; removing worms from, 5, 80–83, 98, 140–141; research on, 54–55, 99, 196–197; tillage of, 63, 67–68, 72, 10; types in Ontario, 31, 47, 49–50

sponsorship program (refugees), 151–152

Statistics Canada, 75

strawberries, 29, 139, 162

substitution: as concept in agrarian economy, 78, 186

subsumption:

—CONJOINED SUBSUMPTION OF LABOR AND NATURE: as a concept, 20–25; dialectical nature of, 25–26, 191–193; utility of concept, 29, 134, 154, 162, 192–194, 199

—FORMAL SUBSUMPTION OF LABOR: and assemblages, 134; description of, 17–18;

and expressions pf autonomy and freedom, 159–162, 175; and the informal economy, 105–106; and the labor process, 32, 138

—FORMAL SUBSUMPTION OF NATURE: in the bait industry, 23–24; description of, 18–19; and forms of rent, 78–79; and the informal economy, 105–106; and the labor process, 138, 154, 158

—REAL SUBSUMPTION OF LABOR: description of, 18; dialectical nature of, 24–26; and forms of rent, 78; and freedom, 161–162; and the informal economy, 105–106; and piece rates, 145; power differentials relating to, 165; in the worm industry, 33

—REAL SUBSUMPTION OF NATURE: in the bait industry, 23–24, 121–123, 181–182; description of, 19–20; and freedom, 161–162

supply management: farmer perspectives on, 98; impact on worm-picking industry, 77, 102, 180; overview of, 94–95; politics of, 180; and relative surplus value, 94

surplus value: absolute forms of, 18, 23–24, 26, 136, 161–162, 195; and assemblages, 28–29, 33, 178, 188; definition of, 15; imperative to accumulate, 38, 178, 190, 192–193, 195, 200; in the informal economy, 125, 131; and nature, 16–20, 194; relationship to subsumption, 17–20, 22–23, 134, 161–162, 191; relative forms of, 18, 94, 121–123, 162, 203n28; and rent, 77, 79

tax: evasion of, 5, 24, 31, 106, 190; fraud, 9, 127–129; income, 112; and the informal economy, 32, 35, 105–106, 131; and subsumption of labor, 106; of worm pickers, 166–168, 198

Temporary Foreign Worker Program (TFWP): conditions of housing, 218n42; difficulty in accessing, 174–175; freedom within, 169–170, 172–174; gender in, 149; overview of, 159; perspectives on, 170–171; as a solution to labor shortages, 166

CRITICAL ENVIRONMENTS: NATURE, SCIENCE, AND POLITICS

Edited by Julie Guthman and Rebecca Lave

The Critical Environments series publishes books that explore the political forms of life and the ecologies that emerge from histories of capitalism, militarism, racism, colonialism, and more.

Founded in 1893,
UNIVERSITY OF CALIFORNIA PRESS
publishes bold, progressive books and journals
on topics in the arts, humanities, social sciences,
and natural sciences—with a focus on social
justice issues—that inspire thought and action
among readers worldwide.

The UC PRESS FOUNDATION
raises funds to uphold the press's vital role
as an independent, nonprofit publisher, and
receives philanthropic support from a wide
range of individuals and institutions—and from
committed readers like you. To learn more, visit
ucpress.edu/supportus.